GENETIC
ENGINEERING
IN THE
PLANT SCIENCES

GENETIC ENGINEERING
IN THE
PLANT SCIENCES

edited by

Nickolas J. Panopoulos

Department of Plant Pathology
University of California
Berkeley, California

PRAEGER

PRAEGER SPECIAL STUDIES • PRAEGER SCIENTIFIC

Library of Congress Cataloging in Publication Data
Main entry under title:

Genetic engineering in the plant sciences.

Includes bibliographies and index.
1. Plant genetic engineering.   I. Panopoulos,
Nickolas J.
QK981.5.G46   581.8'7322   81-10564
ISBN 0-03-057026-3   AACR2

44, 318

Published in 1981 by Praeger Publishers
CBS Educational and Professional Publishing
a Division of CBS Inc.
521 Fifth Avenue, New York, New York 10175 U.S.A.

© 1981 by Praeger Publishers

123456789   145   987654321

Printed in the United States of America

CONTRIBUTORS———————————

Frederick M. Ausubel
Cellular and Developmental
  Biology Group
Department of Biology
The Biological Laboratories
Harvard University
Cambridge, Massachusetts

Susan E. Brown
Cellular and Developmental
  Biology Group
Department of Biology
The Biological Laboratories
Harvard University
Cambridge, Massachusetts

S. D. Daubert
Department of Plant Pathology
University of California
Davis, California

Gary Ditta
Department of Biology
University of California-San Diego
La Jolla, California

D. Dudits
Institute of Genetics
Biological Research Centre
Hungarian Academy of Sciences
Szeged, Hungary

Robert M. Faust
U.S. Department of Agriculture
Insect Pathology Laboratory
Beltsville, Maryland

G. Feix
Institute für Biologie
Albert-Ludwigs Universität
Freiburg, West Germany

R. Gardner
Department of Plant Pathology
University of California
Davis, California

K. Gausing
Department of Molecular Biology
University of Aarhus
Aarhus, Denmark

Robert M. Goodman
Department of Plant Pathology
University of Illinois
Urbana, Illinois

Milton P. Gordon
Department of Biochemistry
University of Washington
Seattle, Washington

B. Gronenborn
Department of Plant Pathology
University of California
Davis, California

Roger Hull
Department of Virology
John Innes Institute
Norwich, England

B. Jochimsen
Department of Molecular Biology
University of Aarhus
Aarhus, Denmark

P. Jørgensen
Department of Molecular Biology
University of Aarhus
Aarhus, Denmark

P. Langridge
Institute für Biologie
Albert-Ludwigs Universität
Freiburg, West Germany

Sharon R. Long
Cellular and Developmental
  Biology Group
Department of Biology
The Biological Laboratories
Harvard University
Cambridge, Massachusetts

K. A. Marcker
Department of Molecular Biology
University of Aarhus
Aarhus, Denmark

Harry M. Meade
Cellular and Developmental
  Biology Group
Department of Biology
The Biological Laboratories
Harvard University
Cambridge, Massachusetts

Lois K. Miller
Department of Bacteriology and
  Biochemistry
University of Idaho
Moscow, Idaho

Kenji Nagahari
Central Research Laboratories
Mitsubishi Chemical Industries
  Limited
Yokohama, Japan

Eugene W. Nester
Department of Microbiology and
  Immunology
University of Washington
Seattle, Washington

Robert C. Nutter
Division of Biological Sciences
Stauffer Chemical Company
Richmond, California

Lowell D. Owens
U.S. Department of Agriculture
Science and Education
  Administration
Cell Culture and Nitrogen Fixation
  Laboratory
Beltsville, Maryland

K. Paludan
Department of Molecular Biology
University of Aarhus
Aarhus, Denmark

Nickolas J. Panopoulos
Department of Plant Pathology
University of California
Berkeley, California

Charles F. Reichelderfer
Department of Entomology
University of Maryland
College Park, Maryland

R. J. Shepherd
Department of Plant Pathology
University of California
Davis, California

Jerry L. Slightom
Department of Chemistry and
  Biochemistry
Southern Illinois University
Carbondale, Illinois

Jeffrey N. Strathern
Cold Spring Harbor Laboratory
Cold Spring Harbor, New York

Michael T. Sung
Department of Chemistry and
  Biochemistry
Southern Illinois University
Carbondale, Illinois

Z. R. Sung
Departments of Plant Pathology
  and Genetics
University of California
Berkeley, California

Michael F. Thomashow
Department of Bacteriology and
  Public Health
Washington State University
Pullman, Washington

Curtis B. Thorne
Department of Microbiology
University of Massachusetts
Amherst, Massachusetts

E. Truelsen
Department of Molecular Biology
University of Aarhus
Aarhus, Denmark

U. Wienand
Institut für Biologie
Albert-Ludwigs Universität
Freiburg, West Germany

FOREWORD ───────────────────────────

The design and construction of better crop plants has always been a goal of plant breeders. As much an art as it is a science, plant breeding has always depended heavily on the empirical methods inherent in the breeder's eye. The ability to identify and select winning combinations of characters early in a breeding program is a most important ingredient in success. The development of cytogenetics and mutation breeding, however, afforded new engineering techniques to shape raw materials that are restricted only by the limits to sexual hybridization and the encoded genetic information already in place. As we enter the 1980s, it seems that even these limits may no longer determine what is possible in effecting genetic change and exchange.

Genetic engineering has taken on a new meaning with the discovery that restriction endonucleases can be used to cleave deoxyribonucleic acid (DNA) at specific sites. The cleaved fragments can be recombined with DNA ligase to create new hybrid molecules that replicate in appropriate host cells. At first restricted to *Escherichia coli* and similar gram-negative bacteria, we have seen the range of host cells extended to include gram-positive bacteria, yeast and *Neurospora*, and animal cells. Plant cells are clearly candidates once general transformation techniques have been developed that extend the range of what is now possible with the Ti-plasmid of *Agrobacterium tumefaciens*. Plant genes that have already been cloned in *E. coli* include complementary DNA (cDNA) made from zein, leghemoglobin and other similar messenger ribonucleic acids (mRNAs), the DNA coding for 5s ribosomal RNA of wheat and rye, chloroplast DNA of maize and wheat, and putative replication origins from maize, among others. Already the challenge is to devise experiments of practical significance for the production of improved crop plants.

These papers, several of which were presented in a symposium organized for the American Phytopathological Society by Dr. Nickolas J. Panopoulos, afford a glimpse of the emerging technology and of the diverse needs and opportunities for its application in the plant sciences. This was an appropriate forum if only because the fear that new plants and new pathogens might run riot was an obsession during the beginnings of genetic manipulation. It is now clear that these concerns were largely unfounded. The years ahead should show how much the benefits will outweigh the risks that haunted those who compiled the National Institutes of Health Guidelines for Recombinant DNA Research in 1976.

Peter R. Day

EDITOR'S NOTE————————————————

Genetic Engineering in the Plant Sciences contains eight chapters (Chapters 1, 7, 8, 9, 10, 12, 13, and 15) that were originally presented at a symposium sponsored by the American Phytopathological Society (APS) as part of its 1979 annual meeting, held concurrently with the Second International Congress of Plant Protection on August 5–13, 1979, in Washington, D.C. The other chapters were subsequently solicited to provide a broader coverage of the rapid developments in genetic manipulations research in fields related, directly and indirectly, to plant sciences and agriculture.

Organization of the symposium and publication of this book would not have been possible without the financial support provided by the National Science Foundation (Grant no. PCM79-16533), the APS, the Department of Agriculture, and the interest expressed by Praeger Scientific in publishing the book. My thanks are also extended to Dr. E. Civerolo for assisting in the organization of the APS symposium, the authors of the various chapters, and the Department of Plant Pathology, University of California-Berkeley, for providing secretarial resources.

I wish to dedicate this book to the memory of my brother Constantine and to all peaceful and constructive uses of recombinant DNA and other genetic manipulations technologies to which this book, hopefully, will contribute.

Nickolas J. Panopoulos

CONTENTS ———————————

1

TOWARD THE GENETIC TRANSFORMATION OF PLANT CELLS

Lowell D. Owens

INTRODUCTION

For this discussion *genetic transformation* is defined as the genotypic alteration of cells caused by the introduction of a deoxyribonucleic acid (DNA)
molecule. This definition acquires a somewhat altered meaning when used
to describe the phenomenon of heritable change in higher plant cells in that
it does not necessarily imply integration of donor DNA into the nuclear or
organelle chromosomes, nor does it imply sexual transmissibility from regenerated individuals. Both issues, however, have bearing on the potential
application of this technique to the improvement of agronomic crops.

The phenomenon of genetic transformation by isolated DNA was first
demonstrated with bacteria some 35 years ago (Avery, MacLeod, and Mc-
Carty 1944). Several decades were to pass before analogous experiments
with plants or cultured plant tissues were reported. However, the potential
usefulness of such a technique became dramatically evident when the totipotency of plant cells—that is, the ability of a single cell to regenerate a
complete plant—was unequivocally demonstrated in 1965 (Vasil and Hildebrandt 1965).

A second major technological advance important to our subject was
Cocking's (1960) discovery of how to prepare large numbers of higher plant
protoplasts by enzymatically digesting away the cell wall. This discovery
was soon followed by reports of the successful culturing of protoplasts of
several higher plants (Cocking 1972). Cell walls were observed to re-form,
sustained cell division occurred, and in some instances, the dividing cells
could be induced to undergo morphogenesis to regenerate the whole plant
(Bhojwani, Evans, and Cocking 1977). Thus, investigations of the 1960
decade succeeded in providing for the first time a form of plant cells that
was amenable to genetic transformation experiments and, further, a basis

1

for the expectation that a transformed cell could become a transformed plant.

Aoki and Takebe (1969) provided the first step in the direction of transformation with their demonstration that tobacco leaf protoplasts could be infected with ribonucleic acid (RNA) isolated from tobacco mosaic virus (TMV). The percentage of infected protoplasts was encouragingly high, and the titer of virulent TMV particles increased logarithmically for 22 hours to a final titer of 10^5 per infected protoplast. Although the biological process occurring was one of transfection, not transformation, the study was, nevertheless, quite important because it unequivocally demonstrated that a naked nucleic acid could survive protoplast uptake and remain fully functional in a biological sense.

The first report of DNA uptake of plant protoplasts was that of Ohyama, Gamborg, and Miller (1972). Their physical studies showed that at least some of the DNA taken up survived for several hours in polymeric form as discerned by its ability to be precipitated by trichloroacetic acid. The question of whether the surviving DNA was sufficiently intact to function biologically was not addressed by this or the many subsequent investigations of a similar nature (Kleinhofs and Behki 1977). Two useful findings that emerged from these studies are that damaged protoplasts and cell wall debris rapidly bind radioactively labeled DNA (Hughes, White, and Smith 1978; Kool and Pelcher 1978) — thereby complicating the interpretation of uptake data — and that the polycation poly-l-ornithine protects bacterial plasmid DNA from degradation by exogenous deoxyribonuclease (DNase) found in protoplast preparations (Hughes, White, and Smith 1979).

The biological fate of DNA taken up by higher plant protoplasts has been the subject of very few reported investigations. The large body of literature describing attempts to transform whole plants, pollen, cultured cells, and callus will not be discussed here because evidence for purported successes remains tentative or ambiguous (Kleinhofs and Behki 1977).

The first detailed report of an attempt to genetically transform higher plant protoplasts was that of Uchimiya and Murashige (1977). They treated protoplasts from a TMV-susceptible isoline of tobacco with DNA purified from a TMV-resistant isoline. Some 500 plants were regenerated from the treated protoplasts and challenged with TMV. None proved to have acquired the resistance trait.

In analyzing their experiment one can appreciate the problems arising from a combination of two factors — the use of donor DNA that inherently has a low concentration of the marker gene and an inadequate screen for large numbers of treated cells. As shown in Table 1.1, in Uchimiya and Murashige's experiment each of 10^6 protoplasts was presented with 1.5 μg of donor DNA. Assuming one copy of the marker gene (TMV resistance) per haploid genome and a haploid genome content of 1 pg per tobacco cell, one

TABLE 1.1
Genetic Transformation Experiments with Mammalian and Plant Cells

Donor DNA		Recipient Cells	DNA Treatment[a]		DNA Uptake		Transformation Frequency	Reference
Source	Size (pg)		Amount per 10^6 Cells	Donor Copies per Cell	Per Cell (pg)	Donor Copies per Cell		
HSV fragment	3.7×10^{-6}	Mouse	40 pg	10	—	—	$1/10^6$	Wigler et al. (1977)
Mouse chromosome	3.3 (haploid)	Mouse	20 µg	6	—	—	$1/10^6$-$10/10^6$	Wigler et al. (1978)
Yeast fragment-plasmid hybrid[b]	8×10^{-6}	Yeast protoplasts	1 µg	120,000	—	—	$1/10^6$-$1/10^3$	Struhl et al. (1977)
Tobacco chromosome	1 (haploid)	Tobacco protoplasts	1.5 µg	1.5	0.05	0.05	0/500	Uchimiya and Murashige (1977)
E. coli plasmid ColE1-kan	1.5×10^{-5}	Tobacco protoplasts	2.5 µg	167,000	0.07	4,800	0/272,000	Owens (1979)

[a]Treatment rates normalized to 10^6 cells.
[b]Recombinant molecule of yeast DNA and E. coli plasmid pBR322.
Source: Compiled by the author.

can calculate from the uptake (0.05 pg per cell) that an average of 1 proto-plast in 20 took up one copy of the marker gene. Further, since only 500 cells were screened — as regenerated plants — for expression of the marker gene, the transformation frequency would have to be on the order of 1 per 500 to be detected.

Transformation frequencies with other eukaryotic systems have not been that high. For example, mouse cells treated with 10 or 6 marker gene copies per cell (Table 1.1: Wigler et al. [1977] and Wigler et al. [1978], respectively) — whether in the form of a small purified restriction fragment of herpes simplex virus (HSV) DNA or unfractionated whole-cell genomic DNA — were transformed with frequencies ranging from 1 per 10^6 to 1 per 10^5. Interestingly, the concentration of the marker gene per unit of donor DNA had little effect on the transformation frequency as long as the amount of donor DNA was adjusted to provide approximately equal num-bers of marker gene copies per recipient cell. The concentration of the marker gene per unit HSV fragment DNA is about a million times that of mouse chromosomal DNA. Could the transformation frequency have been greatly increased by using many more marker gene copies per cell? Al-though a dose response was obtained with nanogram amounts of HSV frag-ment DNA per 10^6 cells (Wigler et al. 1977), the appropriate experiments utilizing the investigators' improved transformation protocol (Wigler et al. 1978) have not been reported. Therefore, the question remains only partial-ly answered for this higher eukaryote.

High doses do produce high transformation frequencies in one lower eukaryote. When yeast cells were treated with a very high number of marker gene molecules, in the form of a cloned recombinant DNA mole-cule, the transformation frequency reached as high as 1 per 1,000 (Table 1.1 Struhl [1979]). These results at least suggest the possibility that transforma-tion frequencies in higher eukaryotes might be dramatically increased by treating cells with many copies of the marker gene.

STUDIES WITH TOBACCO PROTOPLASTS

We have recently attempted to genetically transform tobacco protoplasts with a cloned gene from *Escherichia coli* (Owens 1979). The product of this gene is an enzyme that detoxifies the antibiotic kanamycin to which tobacco cells are sensitive. The gene is borne on a small recombinant DNA mole-cule, ColE1-*kan* (Hershfield et al. 1974), of which it constitutes about a tenth of the entire molecule. The protoplasts we used were derived from tobacco cells cultured in liquid suspension. Initial studies with labeled ColE1-*kan* DNA demonstrated that the protoplasts rapidly bound DNA (Fig. 1.1) but most remained accessible to added DNase. However, some

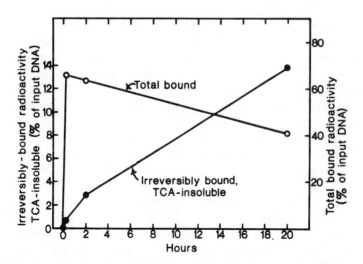

FIG. 1.1. Kinetics of [³H]-ColE1-*kan* DNA binding by tobacco protoplasts. "Total bound" radioactivity represents total cpm associated with protoplasts prior to treatment with DNase; "irreversibly bound," that remaining after DNase treatment.

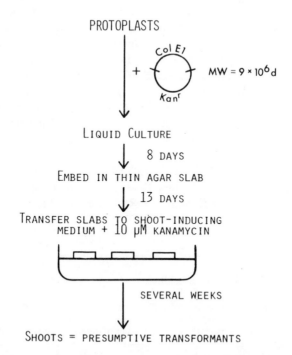

FIG. 1.2. Protocol for genetic transformation experiment.

3 percent became irreversibly bound during 2 hours and could be recovered in polymeric form. This "uptake" represented about 0.07 pg DNA per protoplast. The treated protoplasts were cultured according to the protocol outlined in Fig. 1.2. After about 3 weeks the agar slabs containing thousands of microcalli were placed on selective medium containing kanamycin and shoot-inducing hormones. Several weeks later most calli had turned brown and appeared moribund, but a number of presumptive transformants were obtained as small shoots from cells treated with ColE1-kan DNA (Table 1.2). Fewer than half as many shoots were obtained from control cells treated with plasmid ColE1 DNA. These plantlets were cultured in vitro until about 5 cm tall, at which time several leaves were removed, cut into small pieces, and tested for kanamycin resistance on the same agar medium as above. The results are summarized in Fig. 1.3. The leaf pieces exhibited varying degrees of tolerance to 16 μM kanamycin, but the distribution of tolerance among plantlets regenerated from protoplasts treated with ColE1-kan DNA did not appear to differ significantly from that exhibited by control plantlets. Apparently, the presumptive transformants were a manifestation of natural variance and not of expression of the ColE1-kan gene specifying kanamycin resistance.

To place this experiment in context of the foregoing discussion, I have summarized it in Table 1.1 (Owens [1979]). The DNA treatment of 2.5 μg per 10^6 protoplasts represented 167,000 marker gene copies per cell. The uptake or irreversible binding of 0.07 pg represented the equivalent of 4,800 copies per cell, although some binding to debris as well as some degradation undoubtedly occurred. Finally, 272,000 microcalli were screened in our unsuccessful search for transformants. Our failure to obtain expression of the plasmid-borne gene in tobacco could be due to any of several reasons: a transformation frequency of less than 1 per 2.7×10^5, biochemical barriers

TABLE 1.2——————————————————————————————————
Summary of Two Experiments Testing for Expression of the ColE1-kan Gene for Kanamycin Resistance in Tobacco Cells

| | DNA Treatment | |
Number	ColE1 (control)	ColE1-kan
Protoplasts treated	8.2×10^6	7.3×10^6
Microcalli screened on kanamycin medium	326,000	272,000
Plantlets obtained	32	52
Plantlets tested further for kanamycin resistance*	22	47

*Plantlets exhibiting gross morphological abnormalities or severe stunting were excluded from further testing.

Source: Compiled by the author.

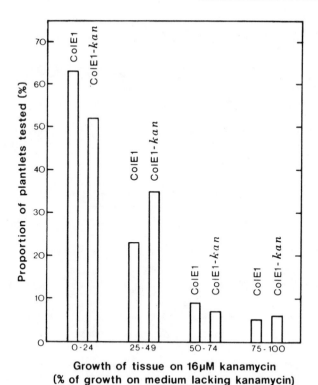

FIG. 1.3. Response to kanamycin of leaf section from plantlets regenerated from protoplasts that were treated with ColE1 (control) or ColE1-*kan* DNA and that survived the initial screen for transformants. Leaf pieces were placed on shoot-inducing agar medium containing 0 or 16 μM kanamycin, and callus and shoot tissues harvested 30 days later. Bars represent the proportion (percent) of plantlets from each treatment that had tissue dry weights within the range indicated (expressed as percent of dry weight obtained on medium lacking kanamycin).

at the level of transcription or translation, or failure to be replicated in the plant cell.

Although our investigation did not resolve the question of whether an *E. coli* plasmid DNA molecule can be maintained and expressed in a plant cell, it does raise the question of whether there are more likely molecules for this purpose. Table 1.3 lists several that are currently considered as potential gene vectors for plant cells. The tumor-inducing plasmid from the crown gall pathogen *Agrobacterium tumefaciens* may be the furthest advanced, since in nature it is known to genetically transform cells of its host plant (Chilton et al. 1978; Schell et al. 1979). Nutter and collaborators have elaborated on this work in Chapter 8.

TABLE 1.3————————————————
Potential Gene Vectors for Higher Plants

DNA Source	Reference
Plant DNA viruses	Hull (1978)
Tumor-inducing plasmid from *Agrobacterium tumefaciens*	Schell et al. (1979)
Chloroplast DNA replication initiation sequence	Bedbrook, Kolodner, and Bogorad (1977)
Transposonlike DNA from mitochondria of cytoplasmic male-sterile maize	Levings and Pring (1979)
Plant DNA replicator sequences	Stinchcomb et al. (1980)

Source: Compiled by the author.

Efforts to transform plant protoplasts with purified preparations of tumor-inducing plasmid DNA are now under way in several laboratories. These efforts may be aided by the rapidly advancing technology of synthetic liposomes. Matthews et al. (1979) and Lurquin (1979) have already demonstrated that nucleic acids can be readily encapsulated into liposomes and efficiently transferred into plant protoplasts.

Given the current state of plant protoplast technology and the availability of suitable DNA molecules, it is reasonable to expect that the genetic transformation of higher plant cells will be conclusively demonstrated within the next year or two. However, the goal of predictably altering the genotype of plants in ways that result in more than a few economically useful phenotypes will require basic knowledge of the processes of genetic transformation and control of gene expression. These tasks will likely occupy plant scientists and molecular geneticists for many years to come.

REFERENCES

Aoki, S., and Takebe, I. 1969. Infection of tobacco mesophyll protoplasts by tobacco mosaic virus ribonucleic acid. *Virology* 39:439–48.

Avery, O. T.; MacLeod, C. M.; and McCarty, M. J. 1944. Studies on the chemical nature of the substance inducing transformation of pneumococcal types. *J. Exp. Med.* 79:137–57.

Bedbrook, J. R.; Kolodner, R.; and Bogorad, L. 1977. Zea mays chloroplast ribosomal genes are part of a 22,000 base pair inverted repeat. *Cell* 11:739–49.

Bhojwani, S. S.; Evans, P. K.; and Cocking, E. C. 1977. Protoplast technology in relation to crop plants: progress and problems. *Euphytica* 26:343–60.

Chilton, M. D.; Drummond, M. H.; Merlo, D. J.; and Sciaky, D. 1978. Highly conserved DNA of Ti-plasmid overlaps T-DNA maintained in plant. *Nature* 275: 147–49.

Cocking, E. C. 1960. A method for the isolation of plant protoplasts and vacuoles. *Nature* 187:962–63.

——. 1972. Plant cell protoplasts – isolation and development. *Ann. Rev. Plant Physiol.* 23:29–50.

Hershfield, V.; Boyer, H. W.; Yanofsky, C.; Lovett, M. A.; and Helinski, D. R. 1974. Plasmid ColE1 as a molecular vehicle for cloning and amplification of DNA. *Proc. Nat. Acad. Sci. (USA)* 71:3445–50.

Hughes, B. G.; White, F. G.; and Smith M. A. 1978. Contribution of damaged protoplasts to DNA uptake by purified plant protoplasts. *Plant Sci. Lett.* 11:199–206.

——. 1979. Fate of bacterial plasmid DNA during uptake by barley and tobacco protoplasts. II. Protection by poly-L-ornithine. *Plant Sci. Lett.* 14:303–10.

Hull, R. 1978. The possible use of plant viral DNAs in genetic manipulation in plants. *Trends Biochem. Sci.* 3:254–56.

Kleinhofs, A., and Behki, R. 1977. Prospects for plant genome modification by nonconventional methods. *Ann. Rev. Genet.* 11:79–101.

Kool, A. J., and Pelcher, L. E. 1978. Conditions that affect binding of DNA by cultured plant cells and protoplasts: an autoradiographic analysis. *Protoplasma* 97:71–84.

Levings, C. S., and Pring, D. R. 1979. Mitochondrial DNA of higher plants and genetic engineering. In *Genetic engineering: principles and methods*, ed. A. Hollaender and J. K. Setlow, pp. 205–22, vol. 1. New York: Plenum Press.

Lurquin, P. F. 1979. Entrapment of plasmid DNA by liposomes and their interaction with plant protoplasts. *Nucl. Acids Res.* 6:3733–84.

Matthews, B. F.; Dray, S.; Widholm, J.; and Ostro, M. 1979. Liposome-mediated transfer of bacterial RNA into carrot protoplasts. *Planta* 145:37–44.

Ohyama, K.; Gamborg, O. L.; and Miller, R. A. 1972. Uptake of exogenous DNA by plant protoplasts. *Can J. Bot.* 50:2077–80.

Owens, L. D. 1979. Binding of ColE1-*kan* plasmid DNA by tobacco protoplasts: nonexpression of plasmid gene. *Plant Physiol.* 63:683–86.

Schell, J.; Van Montagu, M.; DeBeuckeleer, M.; DeBlock, M.; Depicker, A.; DeWilde, M.; Engler, G.; Genetello, C.; Hernalsteens, J. P.; Holsters, M.; Seurinck, J.; Silva, B.; van Vliet, F.; and Villarroel, R. (1979). Interactions and DNA transfer between *Agrobacterium tumefaciens*, the Ti-plasmid and the plant host. *Proc. Royal Soc. Lond. B.* 204:251–66.

Stinchcomb, D. T.; Thomas, M.; Kelly, J.; Selker, E.; and Davis, R. W. 1980. Eucaryotic DNA segments capable of autonomous replication in yeast. *Proc. Nat. Acad. Sci. (USA)* 77:4559–63.

Struhl, K.; Stinchcomb, D. T.; Scherer, S.; and Davis, R. W. 1979. High-frequency transformation of yeast: autonomous replication of hybrid DNA molecules. *Proc. Nat. Acad. Sci. (USA)* 76:1035–39.

Uchimiya, H., and Murashige, T. 1977. Quantitative analysis of the fate of exogenous DNA in *Nicotiana* protoplasts. *Plant Physiol.* 59:301–8.

Vasil, S., and Hildebrandt, A. C. 1965. Differentiation of tobacco plants from single, isolated tobacco cells in microcultures. *Science* 150:889–92.

Wigler, M.; Pellicer, A.; Silverstein, S.; and Axel, R. 1978. Biochemical transfer of single-copy eucaryote genes using total cellular DNA as donor. *Cell* 14: 725–31.

Wigler, M.; Silverstein, S.; Lee, L.-S.; Pellicer, A.; Chang, Y.; and Axel, R. 1977. Transfer of purified herpes virus thymidine kinase to cultured mouse cells. *Cell* 11:223–32.

2

CARROT SOMATIC CELL GENETICS

Z. R. Sung
D. Dudits

INTRODUCTION

Since Carlson's report on the isolation of mutants and production of somatic hybrids in plants from tissue culture (Carlson 1970, 1973; Carlson, Deering, and Smith 1972), there has been strong interest in the establishment of plant somatic cell genetic systems. In the past 10 years biochemical mutants and somatic cell hybrids, interspecific as well as intergeneric, have been obtained in several plant species (see Maliga 1978; Vasil, Ahuja, and Vasil 1979, for review). This chapter summarizes the present state of somatic genetics in carrot with emphasis on those attributes that make it a suitable somatic cell system for genetic studies.

A model somatic system must be easily amenable to a variety of manipulations in culture. The desirable attributes include: availability of haploid cell lines for the isolation of isogenic mutant strains; ability to regenerate plants from cultured cells, protoplasts, and somatic hybrids for sexual genetic studies and for study of gene expression in differentiated organs and tissues; and fast growth as friable liquid cultures for the convenience in mutant isolation and subsequent characterization. Several plant cell culture systems possess these attributes—that is, tobacco, *Datura*, *Petunia*, potato, and carrot, among others.

With reference to carrot, it is well known that both the wild carrot (Queen Anne's lace) and the domesticated carrot produce friable liquid suspension cultures. Most callus cultures initiated from the carrot plants possess efficient and lasting ability to regenerate (Halperin and Wetherell 1964; Reinert 1959; Steward, Mapes, and Smith 1958). Regeneration of plants

The research reported was supported in part by USDA Competitive Grant no. AG5901-0410-9-0370-0.

takes place via embryogenesis. Pure and complete embryo culture can be obtained (Sung, Horowitz, and Smith 1979). Production of protoplasts and protoplast fusion are easily accomplished (Dudits et al. 1979). A number of male-sterile lines useful for genetic hybridization studies are available (Straub 1971). Recently, a number of variants have been isolated and some have been characterized via somatic hybridization (discussed later in this chapter). In addition, somatic segregation, useful for the characterization of recessive markers, has been demonstrated in somatic hybrids of carrot (Lazar et al. 1979). Techniques have also been developed to isolate chromosomes and to enclose them in vesicles. Introduction of isolated plant chromosomes via the vesicles into protoplasts provides a potentially powerful system for the transfer of foreign genetic material (Szabados, Hadlaczky, and Dudits 1981). The carrot somatic cell system has been employed in a variety of studies, including host-pathogen interaction (Jacques and Sung 1978,1981), gene expression of differentiated functions (Sung and Okimoto, in press), developmental genetics of embryogeny (Breton and Sung, submitted), and regulation of metabolite synthesis and nutrient uptake (Furner, unpubl.; Widholm 1974).

This chapter is not intended to be a thorough review of carrot tissue culture; rather it attempts to give a general picture of the present state of carrot somatic genetics and its use for investigating the genetic mechanisms underlying growth and differentiation.

CYTOLOGY AND GROWTH RATES OF CARROT CULTURES

Carrot is a true diploid with $2n = 18$. The haploid (monoploid) chromosome number of 9 gives a comparatively simple karyotype. The chromosomes are to some extent distinguishable. Bayliss (1975b) has divided the nine pairs of chromosomes into one pair of satellited centromeres, four pairs with submedian centromeres, and four pairs with median centromeres.

Carrot cultures are usually maintained in defined medium (Murashige and Skoog 1962). The mass doubling time of carrot suspension and callus cultures can be conveniently recorded by the sidearm-turbidity method (Sung 1976b; Sung and Jacques 1980). It varies from 2 to 6 days, depending on the cell line and culture conditions. It is often possible to select for fast-growing lines from an established slow-growing culture. The cell cycle, measured by continuous labeling with [^{3}H]-thymidine, corresponds well with the doubling time (Bayliss 1975a). In the presence of 2,4-dichlorophenoxyacetic acid (2,4-D), the culture produces embryos with a mean cell cycle time of 33 hours (Bayliss 1975a). The duration of the cell cycle in root tips was found to be 7.5 hours. While the duration of the S phase in culture was similar to that in root tips, the duration of the mitotic phase was different, largely because of differences in the duration of the G_1 phase.

Daucus carota L., a member of the Umbelliferae, has small flowers that present a problem for anther culture, but it should be amenable to pollen culture. To date the only haploid culture was initiated from a haploid carrot plant in Straub's laboratory by Andrea Trockenbroadt (J. Straub 1975: personal communication). The haploid plant was isolated by the twin-seedling method from a domesticated carrot, *Daucus carota* Juwarot. Straub's groups showed that the frequency of haploid plants was very low. Twin seedlings appeared with a frequency of 2.6×10^{-3}. Examination of the metaphase of 600 twin seedlings showed only one plant to be haploid.

A suspension culture of the haploid callus was initiated in 1977. Its karotype has been monitored regularly (Smith, Furner, and Sung, in press). The percentage of haploid mitoses of this culture remained remarkably stable. After 3 years of continuous propagation in suspension culture, the line consisted of 80 percent haploid cells, the remainder being diploids, tetraploids, and aneuploids.

MUTANT ISOLATION

Resistant Variants

The isolation of antimetabolite-resistant mutants in carrot and other plant cell lines has been reviewed (Maliga 1978; Widholm 1977b). The selection for resistant variants is simple and direct. They have been isolated either from liquid batch cultures (Widholm 1972) after long periods of incubation in antimetabolite-supplemented media or from agar plates as individual surviving colonies. The list of resistant carrot lines includes variants resistant to 5-methyltryptophan (Sung 1976a,1976b,1979; Widholm 1972,1974); *p*-fluorophenylalanine (PFP[r]) (Palmer and Widholm 1975); S(2-aminoethyl)-L-cysteine (AEC[r]) (Matthews, Shye, and Widholm 1980); α-aminocaprylic acid, selenomethionine, hydroxyproline, ethionine (Eth[r]) (Widholm 1977b); 2,4-dichlorophenoxyacetic acid (2,4-D[r]) (Breton and Sung, submitted; Widholm 1977b); colchicine (Maliga 1978); 5-fluorouracil (5-FU[r]) (Jacques and Sung 1981; Sung and Jacques 1980); cycloheximide (CH[r]) (Sung 1976a; Sung, Lazar, and Dudits, in press); selenate, chromate, selenocystine (Furner, unpubl.); and lines resistant to multiple amino acid analogues, PFP[r], Eth[r], AEC[r], and 5-MT[r] (Widholm 1976). The hydroxyproline-resistant line is also resistant to L-azetidine-2-carboxylic acid and 3,4-dehydro-DL-proline (Widholm 1976).

If the above traits represent genetic mutations—for example, base substitutions—they might be expected to be stable in the absence of selection, reexpressed in callus from regenerated plants and transmitted to progenies in sexual crosses after meiosis. In most lines listed above, the traits have been shown to be stable for a long period of cultivation in the absence of selective pressure. However, calli, initiated from regenerated plantlets, have in some cases been observed not to be uniformly resistant (Lawyer, Berlyn, and Zetlich 1980). This phenomenon is discussed in a later section.

To date none of these resistance traits have been examined in sexual crosses. This is largely due to practical difficulties. For example, cultivated carrot is biannual; carrot flowers are small; both self- and cross-fertilization occur simultaneously in an umbel. Some lines were isolated from cultures that did not regenerate; others were isolated from a wild carrot line that did not set seeds easily. Variant lines frequently give rise to abnormal plants. However, most of the new lines isolated from the haploid carrot culture develop normally. The difficulties described above can be circumvented by gentle culture manipulations and by crosses to male-sterile plants. Male-sterile lines exist in the same carrot variety *D. carota* Juwarot (Straub 1971, 1978). It is worth noting that morphological abnormalities were often corrected in somatic hybrids. For example, fusion between a *Nicotiana silvestris* and a *N. knightiana* cell line, neither of which could regenerate, produced hybrids that regenerated (Maliga et al. 1977). This phenomenon was also observed in the CH^r carrot line described below.

It is also possible that the traits described above represented stable epigenetic changes, namely, alteration in gene activities, or transposition phenomena involving insertion or excision of deoxyribonucleic acid (DNA) that are potentially reversible after meiosis, as in the case of Ti-DNA transformation in crown gall (Braun and Wood 1976). The criterion of a genetic mutation is that it is heritable. However, this definition now faces new challenges. Meins and Lutz (1980) have observed that an epigenetic change —that is, cytokinin habituation—can be sexually propagated through meiosis to the progeny.

Chlorate resistance has been successfully employed for the isolation of nitrate reductase-deficient mutants in tobacco (Müller and Grafe 1978). A strategy similar to chlorate resistance has been employed to select for carrot strains with altered ability to assimilate sulfate. Strains resistant to chromate, selenate, and selenocystine have been isolated. The characterization of one of the selenate-resistant lines showed that its sulfate uptake system is hypersuppressible by cystine (Ian Furner 1980: personal communication). In the rest of this section, we will discuss some of the better characterized variants with respect to their physiological, biochemical, and genetic characters.

5-Methyltryptophan-Resistant (5-MT[r]) Carrot Lines

Selection of spontaneous 5-MT[r] cell lines usually gives reproducible results. The 5-MT[r] variants can be isolated repeatedly from carrot culture at a frequency of 10^{-6} (Sung, unpubl.) or 3×10^{-7} (Widholm 1977b). (Mutations for other resistance traits occur less frequently and selection is less reproducible.) Many cell lines resistant to 5-MT[r] have been found both in diploid cultures (Sung 1979; Widholm 1972) and in haploid cultures (Sung, unpubl.). The 5-MT[r] cells usually possess an anthranilate synthetase that is less sensitive to inhibition by 5-MT or tryptophan. They often accumulate tryptophan and are auxin autotrophic (Widholm, 1977a). One of the 5-MT[r] autotrophic lines also accumulates indoleacetic acid (IAA) (Sung 1979).

5-MT[r] resistance was found to behave as a dominant trait: somatic hybrids of 5-MT[r] and 5-MT[s] cells were resistant to 5-MT (Harms, Potrykus, and Widholm, in press; Dudits and Sung, unpubl.).

5-Fluorouracil-Resistant (5-FU[r]) Cell Lines

In the course of a mutagenic study (Sung 1976a), three 5-FU[r] cell lines were isolated from a wild type line, W001C. One of them, F5, was characterized further. We found that when arginine biosynthesis is inhibited, W001C cells become more resistant to 5-FU (Jacques and Sung 1978,1981). This result revealed the intimate regulation of the arginine and pyrimidine metabolism. F5 is not only more resistant to 5-FU but also to 5-fluorouridine (Sung and Jacques 1980). However, it is more sensitive to phaseolotoxin, an inhibitor of ornithine transcarbamylase, and has low endogenous arginine levels. Both aspartate transcarbamylase and ornithine transcarbamylase were found to be normal in F5. Because of the metabolite and enzyme compartmentation problem, it was speculated that F5 is altered in the availability of carbamyl phosphate to ornithine transcarbamylase, thus resulting in an excess of pyrimidine and reduced level of arginine.

S(2-Aminoethyl)-L-Cysteine-Resistant (AEC[r]) Cell Lines

An AEC[r] carrot line was found to take up less lysine from the medium than did AEC[s] cells (Matthews, Shye, and Widholm 1980). The aspartokinase activity was reduced in AEC[r] cells, apparently because of the lowered amount of the lysine-sensitive form of the enzyme. This was thought to have been counterbalanced by an increase in the amount of dihydrodipicolinic acid synthetase, the branchpoint control enzyme for the synthesis of lysine. Unlike many other amino acid analogue-resistant cell lines that accumulate certain amino acids — for example, 5-MT[r] and PFP[r] cell lines — AEC[r] cells do not accumulate lysine.

Somatic hybridization studies showed that AEC[r] behaves as a dominant trait: fusion between a 5-MT[r] and an AEC[r] cell line gave rise to hybrids resistant to both amino acid analogues (Harms, Potrykus, and Widholm, in press).

Cycloheximide-Resistant (CH^r) Cell Lines

Stable CH^r cells lines have been isolated from diploid carrot cultures at a frequency of 5.4×10^{-8}. Ethyl methane sulfonate and nitrosoguanidine can increase the frequency 10- to 100-fold. It should be noted that these are frequencies, not rates, since generation time was not taken into consideration in the calculations. The mechanism of resistance is due to CH-inactivation (Sung, Lazar, and Dudits, in press). The trait is expressed in somatic embryos and plantlets but not in the callus of wild type cells. In contrast, both the callus and the embryos of the CH^r cell lines can inactivate CH. Since wild type cells already possess the ability to inactivate CH, CH^r cells differ from the wild type in the regulation of the expression of CH^r. CH^r was shown to be a recessive trait in somatic hybrids (see the section on "Protoplast Fusion").

Temperature-Sensitive Variants

Indirect selection schemes are employed in the isolation of variants displaying a growth disadvantage in relation to the wild type cells. Amino acid auxotrophs and temperature-sensitive growth mutants have been isolated by the BUdR-light (bromouridine deoxyribose) killing method (Carlson 1970; Malmberg 1979). Preliminary experiments indicated that the BUdR-light killing method cannot be employed in carrot tissue culture because following a BUdR treatment, light could not cause killing of the carrot cells (Sung, unpubl.). A filtration-enrichment method has been employed to isolate auxotrophs in moss with limited success (Ashton and Cove 1977).

A new filtration-enrichment procedure was devised, based on the somatic embryogenesis of carrot cells in liquid suspension culture, to isolate temperature-sensitive variants (Breton and Sung, submitted). Employing this selective strategy, we grew a haploid carrot culture, *HA*, in embryogenic medium at 32°C for 3 weeks. The culture was filtered through 100 μ nylon mesh to remove the embryos. The enriched fraction was plated in callus medium at 24°C. The colonies that arose were screened for their ability to grow in both media at two temperatures. Of a total of 285 screened, we found habituated cell lines, variants incapable of callus growth at 32°C (*ts-growth$^-$*); variants capable of callus growth but not embryo development at 32°C (*ts-emb$^-$*); and variants altered in their response to hormones (2,4-D^r). Thus, the above selective scheme is useful for isolating variants of different phenotypes. For example, some of the *ts-growth$^-$* variants may be *ts*-auxotrophs. A similar scheme has been employed to isolate cold-sensitive variants (Lee and Sung, unpubl.).

One of the *ts-emb⁻*, *ts-59*, has been studied further. In the presence of 2,4-D, *ts-59* grew at the same rate as *HA* at both permissive and restrictive temperatures. In the absence of 2,4-D it was capable of normal embryogenesis at 24°C, but at 32°C it grew as callus, occasionally giving rise to globular embryos. Such globular embryos did not develop into heart-shaped or torpedo-shaped embryos. Cells in the globular embryo divided in a disorganized manner (You and Sung, unpubl.). Calli initiated from different plantlets were tested for their ability to form embryos at 32°C and 24°C. All seven regenerated calli behaved like the parents *ts-59*. Control calli, regenerated from four different plantlets of *HA*, were capable of normal embryogenesis at 32°C.

Colored Variants

Orange and purple cell lines often arise spontaneously in carrot cultures (Nishi et al. 1974). The expression of color is very variable in callus culture (Dougall, Johnson, and Whitten 1980; Kinnersley and Dougall 1980). Cell lines expressing color in culture do not usually produce colored embryos or plantlets and vice versa (Sung, unpubl.). Mok, Gabelman, and Skoog (1976) found that cultured cell lines isolated from the same plant exhibit different carotenoid contents, but the phenotypes of plants regenerated from these cell lines were the same. We have found purple calli that give rise to plantlets of normal color, and we have also obtained purple plantlets from yellow calli (Sung, unpubl.). One cycloheximide-resistant (CH^r) line, WCH040, grows as yellow callus but gives rise to plantlets with purple shoots and cotyledons but white roots.

Albino mutants are most useful for somatic hybridization (see next section). The mutant used in the characterization of the CH^r trait, A_1, was isolated as an albino carrot from the M2 generation of a population of mutagenized carrot seeds (Dudits et al. 1979). Because of its recessive, non-reverting nature (frequency of approximately 2×10^{-9}), it has been used universally for genetic characterization of unknown markers by somatic hybridization. In culture albino plantlets have been frequently observed in the haploid line, *HA*, but never in diploid carrot lines. Their frequency was also increased by ethyl methane sulfonate (EMS) mutagenesis (Furner, unpubl.). This finding is consistent with the general understanding that nuclear albinism is usually a recessive trait.

TRANSFER OF GENETIC INFORMATION USING CARROT PROTOPLASTS

Removal of the cell wall from cells grown as a plant organ or as suspension cultures by enzymatic degradation releases protoplasts. Such protoplasts provide a unique system for genetic manipulation of higher plants. In recent years several reviews summarized the trends and results in genetic research using plant protoplasts (Butenko 1979; Cocking 1977; Cove 1979; Melchers 1977; Schieder and Vasil 1980). Regeneration of carrot plants from protoplasts has been achieved both via callus formation (Grambow et al. 1972) and via somatic embryo development (Dudits et al. 1976a).

Protoplast Fusion

The frequency of spontaneous fusion between freshly isolated carrot protoplasts is low, partly because of their small size. Therefore, production of fused cells in sufficient number requires the use of fusogenic agents. Polyethlene glycol (PEG) treatment (Kao and Michayluk 1974; Wallin, Glimelius, and Eriksson 1974) was found to be effective in inducing homo- and heterokaryon fusion of carrot protoplasts (Dudits et al. 1976a,1976b; Reinert and Gosch, 1976). PEG treatment was also successfully applied to induce fusion between carrot protoplasts and HeLa cells to produce interkingdom heterokaryons (Dudits et al. 1976). Carrot protoplasts were also fused with *Xenopus* oocytes at high frequency by Ca^{2+} treatment (Ward et al. 1979). Significant increase in efficiency of PEG-induced fusion was achieved through the use of combined PEG-DMSO (dimethyl sulfoxide) treatment (Haydu, Lazar, and Dudits 1977). The frequency of heteroplasmic fusion between carrot and barley protoplasts was 4 percent for PEG-treated cultures; addition of PEG-DMSO increased the frequency to 15 percent.

As a consequence of membrane fusion during agglutination of protoplasts, nuclei and cytoplasmic organelles from different species are incorporated into a single fusion product. When carrot cell culture protoplasts are fused with leaf protoplasts, mixing of the cytoplasms can be followed visually (Dudits et al. 1976b). Formation of barley-carrot heterokaryons was cytologically detected by differential staining of the nuclei. Detailed analysis of the plant heterokaryons revealed synchronous as well as asynchronous mitosis in the nuclei of the fused cells. Fusion between interphase nuclei also occurred at considerable frequency (Constabel et al. 1975; Dudits et al. 1976b). At present the fate of the fused interphase nuclei is not known; probably they cannot enter into further divisions. The interaction between the fusion partners is significantly influenced by the actual cell cycle stage of the parental cells. Premature chromosome condensation (PCC) in interphase nuclei following fusion with mitotic protoplasts was recently demonstrated in somatic plant cells (Szabados and Dudits 1980). These studies provided additional evidence for mitotic activation during fusion between dividing and nondividing protoplasts, as previously observed in fusion experiments with carrot (Dudits et al. 1976b) and other species (Kao et al. 1974).

Under appropriate culture conditions the fused cells are capable of subsequent divisions and formation of multicellular colonies in a large number of fused cell combinations (Constabel et al. 1976; Dudits et al. 1976b; Dudits, Hadlaczky, Lazar,and Haydu 1980; Gosch and Reinert 1976,1978; Kao and Michayluk 1974; Kao et al. 1974). Some of the intergeneric hybrid colonies grew as nondifferentiated calli of hybrid cell lines (Binding and Nehls 1978; Kao 1977) but often can be induced to differentiate into organs and plants (see "Applications").

Genetic Characterization of Variants by Somatic Hybridization

While genetic characterization of the variants can be performed through sexual crosses, somatic hybridization is the only means of studying those variants that cannot differentiate normally. These problems may result from pleiotropic effects of the mutations or severe selective pressure. Four drug-resistant carrot lines have been characterized by somatic hybridization. 5-MT^r, 5-FU^r, and AEC^r were found to be dominant phenotypes: fusion between a 5-MT^r and an AEC^r carrot line indeed generated double-resistant colonies on the selective medium (containing both 5-MT and AEC). The somatic hybrids were predominantly tetraploid cells (Harms, Potrykus, and Widholm, in press).

In the case of CH^r, direct selection for CH^r fusion products failed to produce hybrids. Subsequent fusion studies employing morphological markers for selection demonstrated that CH^r is a recessive trait. Somatic hybrids constructed between the CH^s cells from a domesticated albino carrot and CH^r cells from a wild carrot line, WCH105, were CH^s. However, the CH^s hybrids segregated CH^r colonies (see "Somatic Segregation"). These findings indicate that CH^r in the WCH105 is a recessive trait (Lazar et al. 1979). The occurrence of somatic segregation proves the trait is stable in the presence of the wild type genome, implying that this is a genetic trait. It should be noted that this result does not eliminate the possibility that CH^r is a cytoplasmic trait. However, a recessive, nuclear-coded albinism marker used in this cross showed similar segregation. Plants regenerated from the CH^r line were incapable of developing beyond the seedling stage, but calli reinitiated from the plantlets were CH^r. Plants regenerated from the hybrid and somatic segregants gave rise to normal vegetative growth but abnormal reproductive organs (Dudits, unpubl.). Therefore, the genetic characterization of the CH^r trait can only be accomplished via somatic hybridization.

Because of the shortage of selectable biochemical markers, somatic hybrids are often selected or screened by using morphological and isozyme markers. X-ray irradiation (Zelcer, Aviv, and Galun 1978) and metabolite inhibitors (Lazar, Dudits, and Sung, in press; Nehls 1978) have been used to eliminate the parents from the fusion mixtures. These methods have been successful but tedious. For the purpose of genetic characterization, new mutations should be isolated from lines already equipped with known genetic markers. However, multiple mutations may generate additional problems such as loss of the regenerative potential of the cell line (see "Embryogenesis in Carrot Culture"). This type of complication can be circumvented by the construction of a line possessing a recessive as well as a dominant trait — for example, a line whose callus is resistant to 5-MT or AEC but produces albino plantlets (Sung, unpubl.). Hybrids between this line and a line carrying a new marker to be characterized can be isolated as 5-MT^r green plantlets. Subsequently, the expression of the unselected marker in the hybrids can be screened to determine whether the mutation is dominant or recessive.

Somatic Segregation

Protoplast fusion brings about the mixing of genetic material, a prerequisite to recombination, and makes genetic complementation and dominance-recessiveness tests possible. Following the mixing of cytoplasms, nuclei often fuse. Mapping and gene linkage studies can be greatly facilitated if genes have been assigned to chromosomes. Somatic segregation involving chromosome elimination has been exploited extensively for the mapping of genes on chromosomes in animal tissue cultures (Ruddle 1973). Chromosome elimination has been observed in interspecific and intergeneric somatic hybrids of plants that contain dimorphic chromosomes (Binding and Nehls 1978; Maliga et al. 1978).

In fusion between two carrot cell lines of the same chromosome morphology, the use of a resistant — therefore selectable — but recessive trait (such as CH^r) has been instrumental in the detection of somatic segregation. Fusion between protoplasts of a CH^r cell line that cannot differentiate normal leaves and an albino line was performed (see "Genetic Characterization of Variants by Somatic Hybridization"; Lazar, Dudits, and Sung, in press). A selective scheme was devised based on the expression of green plantlets capable of differentiating normal dissected leaves. All calli from somatic hybrids were CH^s. However, following prolonged incubation, both albino plantlets and CH^r calli appeared at a frequency of 10^{-4}, 1,000 times more frequent than the appearance of CH^r in the original diploid cell line. This demonstrates the utility of somatic segregation for the genetic analysis of recessive traits in plant tissue culture. Furthermore, the appearance of somatic segregants was accompanied by increased chromosome instability, which suggests that chromosome elimination may have occurred. If the technical capability of karyotyping carrot chromosome is improved, it would be possible to localize the CH^r trait on the chromosome.

Uptake of Organelles, Chromosomes, and DNA

Genetic exchange between somatic cells has been most commonly accomplished via protoplast fusion. However, uptake of genetic material by protoplasts provide a means for introducing more specific genetic information into plant cells.

Isolated plant protoplasts take up a variety of particles including chloroplasts (Davey, Frearson, and Power 1976; Potrykus 1973) and nuclei (Lörz and Potrykus 1978). Uptake of isolated chloroplasts from the alga *Vaucheria dichotoma* into carrot protoplasts was induced by PEG treatment (Bonnett and Eriksson 1974). Even though these membrane-surrounded cell components (nuclei or chloroplasts) were taken up by the protoplasts, they were not successfully (stably) maintained or expressed.

Purified metaphase chromosomes are extensively used as vectors for transfer of genetic information in mammalian cell genetics (reviewed by Willecke [1978]). Conceivably, chromosomes can be sorted such that only one kind of chromosome is used to transform cells. The results from studies on metaphase chromosome-mediated gene transfer with mammalian cells have encouraged the development of basic methods for isolation and transfer of plant chromosomes. The mass isolation of plant chromosomes from mitotic protoplasts has been recently described for various plant species (Malmberg and Griesbach 1980; Szabados, Hadlaczky, and Dudits 1981). The purified plant chromosomes could be taken up by plant protoplasts after PEG treatment as shown by cytological analysis (Szabados, Hadlaczky, and Dudits 1981). Chromosomes from a 5-MT^r carrot line were incubated with the protoplasts of a 5-MT^s albino carrot in the presence of PEG. 5-MT^r colonies were isolated from the 5-MT^s protoplasts. Because of the low frequency of appearance of the 5-MT^r albino colonies, further confirmation of the origin of the 5-MT^r trait is needed. Experiments are in progress to increase the efficiency of the uptake by using liposomes as mediators (Dudits, unpubl.). If these basic techniques can be established, transfer of isolated plant chromosomes may provide new impetus to genetic manipulation of higher plants.

Plant protoplasts have proved to be an appropriate system for incorporation of isolated DNA into plant cells. (Lurquin and Marton 1980; Ohyama, Pelcher, and Schaefer 1978). Carrot protoplasts, among others, were fed radioactive *Escherichia coli* DNA, and about 0.6 to 2.8 percent of this exogenous DNA was taken up (Ohyama, Gamborg, and Miller 1972).

Uptake of DNA could be enhanced by incubation of protoplasts and nucleic acids in the presence of polycations (Ohyama et al. 1972; Suzuki and Takebe 1976), PEG, or Ca^{2+} ions at pH 10 (Lurquin 1980). Protamine sulfate with Zn^{2+} significantly increased the uptake of pBR313 plasmid DNA into carrot protoplasts (Fernandez, Lurquin, and Kado 1978).

The uptake experiments with radioactive single-stranded or double-stranded DNAs indicate that large portions of the absorbed DNA are degraded (Suzuki and Takebe 1976; Uchimiya and Murashige 1977). Therefore, the use of artificial lipid vesicles for entrapment of DNA may provide an improved method of transferring plasmids into protoplasts (Lurquin 1979). Rollo et al. (1980) found that positively charged multilamellar liposomes promoted association of native *Bacillus subtilis* [^{3}H]-DNA with carrot protoplasts with the highest efficiency.

Plant transformation experiments can be facilitated by employing protoplasts prepared from a nonrevertible mutant and recombinant DNA technology. The latter enables the production of a large quantity of DNA (of a specific gene), which preferably contains a plant replicon. Alternatively, the Ti-plasmid from *Agrobacterium tumefaciens* can be used to transfer foreign DNA into plant protoplasts (Davey, Cocking, Freeman, Draper, Pearce, Tudor, Hernalsteens, de Benckeleer, Van Montagu, and Schell, 1980; Davey, Cocking, Freeman, Pearce, and Tudor, 1980; Schilperoort et al. 1978).

Applications

The widespread interest in fusion of plant protoplasts lies largely in the hope of generating new plant species. Somatic hybridization between sexually incompatible and distantly related species offers the opportunity of generating plants of new genetic combinations. The creation of Pomato (Melchers, Sacristan, and Holder 1978) and Arabidobrassica (Gleba and Hoffman 1980) plants demonstrated the feasibility of this approach. This section discusses the results of the production of parasexual hybrids between carrot and other Umbelliferae species. We hope the information presents a general view of the potential and limitations of protoplast fusion technology.

Interspecific Hybridization
Chlorophyll-deficient mutants are widely used for selection of somatic hybrid plants. The normal green phenotype can be restored through genetic complementation between different mutants (Melchers and Labib 1974; Schieder 1977) or through correction of the defect in the photosynthetic apparatus by the wild genotype (Cocking et al. 1977; Schieder 1978). The lat-

ter method was applied in the selection of somatic hybrids between *Daucus carota* (2n = 18) and *D. capillifolius* (2n = 18) (Dudits et al. 1977). After fusion of protoplasts from carrot cells carrying a nuclear albino mutation with green *D. capillifolius*, somatic hybrids were regenerated with intermediate leaf morphology between the parental species. The selected somatic hybrids developed white roots. This finding showed the dominance of the white allele carried by the *D. capillifolius* genome. Cytological studies revealed the doubling of parental chromosomes in most of the somatic hybrids; however, aneuploid plants with 35 and 34 chromosomes were also found among them. After pollination the somatic hybrids produced seeds and the second generation showed segregation for the albino phenotype and for leaf shape. By analogy with other somatic hybrids in Solanaceae, the analysis of *Daucus* interspecific hybrids provided evidence for parasexual integration of complete genomes in related species. The production of amphidiploids between related species by somatic hybridization may contribute to plant breeding by overcoming species incompatibility at the level of pollination, by increasing the genetic stability of hybrids, and by eliminating meiotic abnormalities.

During fusion of plant protoplasts, the cytoplasmic organelles from both of the parents are transferred into the fused cells and somatic hybrids. Biparental transmission of these organelles opened entirely new fields in studies on extrachromosomal inheritance such as chloroplast function and segregation (Belliard et al. 1978; Chen, Wildman, and Smith 1977; Evans, Wetter, and Gamborg 1980; Gleba 1978; Maliga et al. 1980; Melchers, Sacristan, and Holder 1978), cytoplasmic male sterility (Izhar and Power 1979; Zelcer et al. 1978), and mitochondrial recombination (Belliard, Vedel, and Pelletier 1979).

Intergeneric Hybridization

In contrast to the success in somatic hybridization between related species, fusion of intergeneric protoplasts often failed to produce hybrid plants (Cocking 1978; Schieder 1977). Fusion experiments with carrot protoplasts indicated that cellular incompatibility is not expressed during growth of the fused callus cells but only during embryo differentiation (Dudits, Hadlaczky, Lazar, and Haydu 1980). These and other studies directed attention to the expression of somatic incompatibility during various developmental stages.

Since it is often difficult to maintain both chromosome complements in the hybrids, introduction of foreign genetic material may be accomplished by transferring only part of the plant genome. Fusion between dividing protoplasts from albino carrot and mitotically inactive leaf protoplasts from *Aegopodium podagraria* resulted in green plants with *Aegopodium*-specific characters but lack of cytologically detectable *Aegopodium* chromosomes (Dudits et al. 1979; Dudits, Hadlaczky, Lazar, and Haydu 1980; Dudits, Koncz, Bajszar, Hadlaczky, Lazar, and Horvath 1980). Somatic cell hybrids with analogous genetic constitution are known in several fusion combinations with mammalian cells (Boyd and Harris 1973; Graves et al. 1979; Rodgers 1979; Schwartz, Coock, and Harris 1971; Tsumamoto, Klein, and Hatanka 1980). Based on fusion experiments with dividing mammalian cells and inactive chick erythrocytes, it was suggested that PCC in interphase nuclei could cause rapid genome elimination; meanwhile, small fragments from the pulverized nuclei could integrate into the recipient genome (Schwartz, Coock, and Harris 1971). The recently available methods for the isolation of mitotic protoplasts from synchronized cultures (Szabados and Dudits 1980) have made it possible to study the effects of interphase mitotic fusion on the genotype of the resulting somatic hybrid plants.

An increased elimination of one parental genome was achieved by irradiation of protoplasts prior to fusion. Analysis of regenerated hybrid plants after fusion between X-ray irradiated leaf protoplasts of *Petroselinum hortense* and cell culture protoplasts of albino carrot revealed the transfer of parsley genes into the genome of carrot (Dudits, Fejer, Hadlaczky, Koncz, Lazar, and Horvath 1980). The selected plants had a chromosome number close to that of carrot.

Protoplast fusion-mediated gene transfer systems involving mitotically inactive or irradiated protoplasts have been successfully developed by using carrot as a model plant. Further studies are needed to improve the efficiency of these methods — for example, by using genetically marked strains.

Polyploidization by Homokaryon Fusion
The significance of polyploid plants in genetic analysis and in plant breeding is well known. The traditional method of polyploidization, based on colchicine treatment, has several disadvantages — for example, high lethality in treated cells, differential growth of chimeric tissues, and low efficiency. Homokaryon fusion offers an improved method for production of polyploid plants in those species in which plants can be regenerated from cultured protoplasts (Dudits et al. 1976a). The high frequency of plants with increased ploidy level permits efficient selection of polyploid plants by cytological screening.

EMBRYOGENESIS IN CARROT CULTURE

The most important feature of plant tissue cultures, namely, their totipotency, was first observed in carrot cultures that regenerated to plants following embryo formation. The origin (McWilliam, Smith, and Street 1974) and morphology of somatic embryos and the nutritional, biochemical, and hormonal influences (Fujimura and Komamine 1975) affecting embryogenesis have been reviewed (Raghavan 1976; Reinert 1968; Street 1977; Thorpe 1980). This section attempts to investigate the molecular mechanisms underlying somatic embryogenesis in carrot culture.

Metabolic Factors Influencing Embryogenesis

Carrot is considered as one of the best model organisms from the standpoint of regenerative potential. Regeneration in carrots proceeds through embryogenesis. Carrot cells maintained in suspension culture for as long as 7 years are still capable of efficient embryogenesis. However, genetically manipulated strains often lose their regenerative potential or exhibit developmental defects. In some cases, the regenerative problems arise independently of the new phenotypes. In other cases, mutations produce pleiotropic effects that affect differentiation — for example, those that affect cellular metabolism and hormonal balance. In a 5-MT^r line, W001, regeneration is suppressed to a frequency of 10^{-5} (Sung, Horowitz, and Smith 1979). The line accumulates IAA (Sung 1979), which apparently suppresses embryogenesis (Sung 1975). Accumulation of IAA in this line is an example of how altered hormonal level can inhibit embryogenesis.

Poor regeneration in the F5 line, a 5-FU^r cell line, is an example of how altered concentration of other metabolites may affect regeneration (Sung, unpubl.). Normally, callus cells of carrots contain more arginine than plantlets: 9 nmole/10^6 cells in callus and 2.6 nmole/10^6 cells in plantlets. However, F5 callus contains only 0.15 nmole/10^6 cells of arginine (Sung and Jacques 1980).

Arginine decarboxylase activity and polyamine concentrations were shown to elevate when carrot cells were transferred into embryogenic medium (Montague, Koppenbrink, and Jaworski 1978). Since polyamines are thought to play a role in early embryogenesis (Fozard et al. 1980; Montague, Koppenbrink, and Jaworski 1978), low levels of arginine in F5 may limit polyamine synthesis, thus affecting regeneration in F5. Those plantlets that did manage to regenerate in F5, however, contained the same level of arginine as the normal plantlets. They could originate either from a small population of 5-FU^s revertants (F5 has been cloned through a protoplast) or from 5-FU^r cells that achieved the necessary physiological conditions for regeneration via alternative pathways. We believe the regeneration process selects for the 5-FU^s cells, because only 8 out of the 50 calli from regenerated plantlets of F5 are still resistant to 5-FU; the rest have become sensitive to 5-FU. If low levels of arginine concentration are actually the cause of poor regeneration in F5, it appears that cells undergoing differentiation require a more stringent set of physiological and biochemical conditions than those undergoing growth only.

The Genetic Program During Embryogenesis

Regeneration involves a series of morphogenetic steps. In carrot culture, the developmental sequence is: callus → globular-shaped embryo → heart-shaped embryo → torpedo-shaped embryo → plantlets; somatic embryo-genetic steps resemble that of zygotic embryogeny (Borthwick 1931). Although it is possible to manipulate embryogenesis by 2,4-D, transfer to fresh medium, and cell density adjustments (Sung and Okimoto, in press) and to speed up development by 2-isopentenyl-adenine (Sung, Horowitz, and Smith 1979), it is not possible to skip steps — for example, to go directly from callus to torpedo-shaped embryos. The sequential transition from one morphological stage to the next must result from the faithful execution of an underlying genetic program. Such a program seems necessary for directing the temporal and spatial expression of selected groups of genes (Britten and Davidson 1969). By identifying the specific genes expressed during normal embryogenesis and by studying their expression in lines impaired in embryogenesis (for example, 5-MT^r and habituated lines) and in variants altered in the control of embryonic functions (for example, CH^r lines and ts-emb^- mutants), we hope to gain insights into the program.

In carrot cultures it takes about 5 to 6 days following subculture into 2,4-D medium for globular embryos to form. However, many biochemical and cellular activities are undergoing changes during this period, as shown by a number of biosynthetic activities that become expressed at different times prior to the sixth day (Dudits, Lazar, and Bajszar 1979; Matsumoto, Gregore, and Reinert 1975; Montague, Armstrong, and Jaworski 1979; Sengupta and Raghavan 1980a,1980b; Sung and Okimoto, in press). Whether these changes are causally related to embryo development is not known. However, their perturbation is likely to disturb embryogenesis. Conversely, impairment in developmental processes will also alter gene products whose expression is restricted to specialized cell types. Thus, these temporally expressed activities serve as landmarks in the timetable of embryogenesis. Some of them are described below in the order in which they appear during the transition from callus to globular embryo.

Ability to Condition the Culture for the Synthesis of Callus-Specific Proteins (C)

Callus culture at high density is capable of conditioning its medium such that a set of callus-specific proteins are made. Embryos do not have this ability (Sung, unpubl.). The synthesis of the callus-specific proteins as well as the embryo-specific proteins (see below) have been studied by autoradiography and two-dimensional gel electrophoresis of the soluble polypeptides extracted from the cells (Sung and Okimoto, in press).

Synthesis of Two Embryonic Proteins (E)

As soon as cells are subcultured into fresh medium with or without 2,4-D (0.1 mg/ml), the synthesis, or elevated synthesis, of two new proteins is detected within 4 hours (Sung and Okimoto, in press). If the cells were kept at low density in 2,4-D-free medium, synthesis of these proteins continues throughout subsequent embryonic stages up to the plantlets. If the cells are kept at low density in medium containing 2,4-D, globular embryos also develop, but they are converted back to callus before heart embryos form. However, the synthesis of the embryonic proteins ceases before this disruption of the embryo development. (*Note:* The synthesis of the embryonic protein also precedes the morphological change, that is, embryo formation.) Cell lines impaired in embryogenesis, such as W001 (Sung, Horowitz, and Smith 1979), and habituated lines, such as W001C/b and F5R82 (Sung, unpubl.), display an altered synthetic pattern for these proteins. These results suggest that the embryonic proteins play a key role in embryogenesis.

Elevated Arginine Decarboxylase Activity

Arginine decarboxylase activity is elevated at the same time and under the same conditions as synthesis of the embryonic proteins is initiated (Montague, Armstrong, and Jaworski 1979; Montague, Koppenbrink, and Jaworski 1978). As mentioned earlier, the elevated activity is apparently responsible for the altered metabolism in arginine and polyamine during embryogenesis and possibly plays a role in embryo development.

Ability to Inactiviate Cycloheximide (CH^i)

As described in an earlier section, callus cells of normal carrots are sensitive to cycloheximide, but after 3 days in the 2,4-D-free medium, they develop an ability to inactivate CH, which renders them cycloheximide resistant (CH^r). This ability (CH^i) is expressed throughout the subsequent embryonic stages. In that section we described a CH^r variant, WCH105, that expresses the CH^i in callus also (Sung, Lazar, and Dudits, in press).

Formation of Globular Embryos (M)

Globular embryos are first detected about 6 days following low-density subculture into fresh medium.

An examination of the variants showed that the expression of several markers — namely C, E, CH^i, and M — appeared to be controlled in a coordinated fashion by the same genetic program that can be manipulated by 2,4-D, fresh medium, and cell density. Variants impaired in one step show altered expression in other steps.

W001.   Synthesis of E is induced by fresh medium. However, continuing synthesis occurs only in low-density cultures in the absence of 2,4-D. W001 is altered in tryptophan synthesis, which leads to an accumulation of endogenous auxin (IAA). The accumulated IAA inhibits embryogenesis (Sung, Horowitz, and Smith 1979) as well as the synthesis of embryonic proteins in the absence of 2,4-D, and fresh medium cannot induce their synthesis in this line (Sung and Okimoto, in press). W001 grown in embryogenic medium does not develop the ability to inactivate CH (Sung, unpubl.).

F5R82.   F5R82 is a habituated cell line. Its embryonic protein synthesis can be induced by fresh medium but cannot be sustained in the absence of 2,4-D.

WCH105 and WCH040.   WCH105 and WCH040 inactivate CH when grown as undifferentiated callus. The callus of these lines cannot condition the cells to produce the callus-specific proteins; rather it synthesizes the embryonic proteins at all times (Okimoto and Sung, unpubl.).

The above information can be integrated to depict the relationship among these temporal steps that occur during embryogenesis. Auxin exerts a negative control on E, CH^i, and M, but if combined with high cell density, it exerts a positive control on C. In W001 high levels of auxin are always present, resulting in a permanent suppression of E, CH^i, and M. Since an alteration in the regulation of CH^i expression also affects the expression of C and E, as seen in WCH105, the simplest explanation for the simultaneous control of the three phenotypes is the existence of a common site X that controls the expression of E, C, and CH^i but not M. An alteration at step X, as in the case of WCH105 and WCH040, results in a reversed pattern of E, C, and CH^i expression; however, it does not change 2,4-D's effect on embryo morphogenesis (M): WCH105 still grows as callus in the presence of 2,4-D.

The steps involved in the induction of E synthesis can be separated from those that sustain its continuing production, because synthesis of E can be induced but not maintained in F5R82. Since embryogenesis is affected in both W001 and F5R82, both processes or, at least, the perpetual synthetic process is a necessary precursor step(s) to embryo formation (M). However, synthesis of E is not the only precursor step to M as evidenced by the phenotype of WCH105.

CONCLUSION

The above information was derived entirely from results obtained from iso-genic lines of a diploid wild carrot line, W001C (Sung, Horowitz, and Smith 1979). Our approach is to identify more temporal genes that are expressed during embryogenesis and to obtain more variants altered in the expression of these genes in order to map all the steps involved during this developmental program.

It is worth noting that regardless of the genetic or epigenetic nature of the traits used, as long as the variants are stable they are useful in studies aimed at understanding the developmental process. The temperature-sensitive lines impaired in embryogenesis were derived from a haploid cultivated carrot, and they are not yet integrated into this scheme.

In addition to studying the temporal gene activities in the ts-emb⁻ variants, we can determine the stage in which the embryonic process is affected by temperature-shift experiments. This information will allow the formulation of a coherent model that may address the question of the temporal sequence and the causality of these gene activities involved during embryogenesis.

In summary the switch from the undifferentiated callus growth to the organized embryo growth must involve an initial step of change in the genetic program; the readout of this program will result in changes in facultative and constitutive "household" functions (Lewin 1980) and in developmental functions. To date nothing is known about the developmental functions. Presumably, they are coded for by genes that eventually bring about morphogenesis. Since the instantaneous concentration of the metabolites can affect differentiation, variants altered in the regulation of embryo development as well as in the metabolic pathways are valuable in elucidating the molecular mechanisms underlying somatic embryogenesis.

REFERENCES

Ashton, N. W., and Cove, D. J. 1977. The isolation and preliminary characterization of auxotrophic and analogue resistant mutants of the moss *Physcomitrella patens*. *Molec. Gen. Genet.* 154:87–95.

Bayliss, M. W. 1975a. The duration of the cell cycle of *Daucus carota* L. *in vivo* and *in vitro*. *Exp. Cell Res.* 92:31–38.

—— . 1975b. The effects of growth *in vitro* on the chromosome complement of *Daucus carota* (L.) suspension cultures. *Chromosoma* 51:401–11.

Belliard, G.; Pelletier, G.; Vedel, F.; and Quatier, F. 1978. Morphological characteristics and chloroplast DNA distribution in different cytoplasmic parasexual hybrids of *Nicotiana tabacum*. *Molec. Gen. Genet.* 165:231–37.

Belliard, G.; Vedel, F.; Pelletier, G. 1979. Mitochondrial recombinations in cyto-plasmic hybrids of *Nicotiana tabacum* by protoplast fusion. *Nature* 281:401–3.

Binding, H., and Nehls, R. 1978. Somatic cell hybridization of *Vicia faba* and *Petunia hybrida*. *Molec. Gen. Genet.* 164:137–43.

Bonnett, H. T., and Eriksson, T. 1974. Transfer of algal chloroplasts of higher plants. *Planta* 120:71–79.

Borthwick, H. A. 1931. Development of macrogametophyte and embryo of *Daucus carota*. *Bot. Gaz.* 92:23–44.

Boyd, J. L., and Harris, H. 1973. Correction of genetic defects in mammalian cells by the input of small amounts of foreign genetic material. *J. Cell Sci.* 13:841–61.

Braun, A. C., and Wood, H. N. 1976. Suppression of the neoplastic state with the acquisition of specialized functions in cells, tissues and organs of crown gall teratomas of tobacco. *Proc. Nat. Acad. Sci. (USA)* 73:496–500.

Breton, A., and Sung, Z. R. Submitted. Temperature-sensitive variants of a haploid carrot line impaired in somatic embryogenesis.

Britten, R. J., and Davidson, E. H. C. 1969. Gene regulation for higher cells: a theory. *Science* 165:349–57.

Butenko, R. G. 1979. Cultivation of isolated protoplasts and hybridization of so-matic cells. *Int. Rev. Cytol.* 59:323–73.

Carlson, P. S. 1970. Induction and isolation of auxotrophic mutants in somatic cell culture of *Nicotiana tabacum*. *Science* 168:487–89.

——— . 1973. Methionine-sulfoximine-resistant mutants of tobacco. *Science* 180: 1366–68.

Carlson, P. S.; Deering, R. D.; and Smith, H. H. 1972. Parasexual interspecific hy-bridization. *Proc. Nat. Acad. Sci. (USA)* 169:2292–94.

Chen, K.; Wildman, S. G.; and Smith, H. H. 1977. Chloroplast DNA distribution in parasexual hybrids as shown by polypeptide composition of fraction I protein. *Proc. Nat. Acad. Sci. (USA)* 74:5109–12.

Cocking, E. C. 1977. Uptake of foreign genetic material by plant protoplasts. *Int. Rev. Cytol.* 48:323–43.

——— . 1978. Selection and somatic hybridization. In *Frontiers of plant tissue culture 1978*, ed. T. A. Thorpe, pp. 151–58. Calgary, Alberta: International Associa-tion of Plant Tissue Culture, University of Calgary.

Cocking, E. C.; George, D.; Price-Jones, M. F.; and Power, J. B. 1977. Selection procedures for the production of interspecific somatic hybrids of *Petunia hybrida* and *Petunia paraodii*. II. Albino complementation selection. *Plant Sci. Lett.* 10:7–12.

Constabel, F.; Dudits, D.; Gamborg, O. L.; and Kao, K. N. 1975. Nuclear fusion in intergeneric heterokaryons. A note. *Can. J. Bot.* 53:2092–95.

Constabel, F.; Weber, G.; Kirkpatrick, J. W.; and Pahl, K. 1976. Cell divi-sion of intergeneric protoplast fusion products. *Z. Pflanzenphysiol.* 79:1–7.

Cove, D. J. 1979. The uses of isolated protoplasts in plant genetics. *Hered-ity* 43:295–314.

Davey, M. R.; Cocking, E. C.; Freeman, J.; Draper, J.; Pearce, N.; Tudor, I.; Hernalsteens, J. P.; de Benckeleer, M.; Van Montagu, M.; and Schell, J. 1980. The use of plant protoplasts for transformation by *Agrobacterium* and isolated plasmids. In *Advances in protoplast research*, ed. L. Frenczy and G. L. Farkas, pp. 425–30. Budapest: Akademiai Kiado.

Davey, M. R.; Cocking, E. C.; Freeman, J.; Pearce, N.; and Tudor, I. 1980. Transformation of *Petunia* protoplasts by isolated *Agrobacterium* plasmids. *Plant Sci. Lett.* 18:307–13.

Davey, M. R.; Frearson, E. M.; and Power, J. B. 1976. Polyethylene glycol-induced transplantation of chloroplasts into protoplasts: an ultrastructural assessment. *Plant Sci. Lett.* 7:7–16.

Dougall, D. K.; Johnson, J. M.; and Whitten, G. H. 1980. A clonal analysis of anthocyanin accumulation by cell cultures of wild carrot. *Planta* 149:292–97.

Dudits, D.; Fejer, O.; Hadlaczky, Gy.; Koncz, Cs.; Lazar, G. B.; and Horvath, G. 1980. Intergeneric gene transfer mediated by plant protoplast fusion. *Molec. Gen. Genet.* 179:283–88.

Dudits, D.; Hadlaczky, Gy.; Bajszar, Gy.; Koncz, Cs.; Lazar, G.; and Horvath, G. 1979. Plant regeneration from intergeneric cell hybrids. *Plant Sci. Lett.* 15:101–12.

Dudits, D.; Hadlaczky, Gy.; Lazar, G.; and Haydu, Zs. 1980. Increase in genetic variability through somatic cell hybridization of distantly related plant species. In *Plant cell cultures: results and perspectives*, ed. F. Sala, B. Parisi, R. Cella, and O. Ciferri, pp. 207–14. Amsterdam: Elsevier/North-Holland.

Dudits, D.; Hadlaczky, Gy.; Levi, E.; Fejer, O.; Haydu, Zs.; and Lazar, G. 1977. Somatic hybridization of *Daucus carota* and *D. capillifolius* by protoplast fusion. *Theor. Appl. Genet.* 51:127–32.

Dudits, D.; Kao, K. N.; Constabel, F.; and Gamborg, O. L. 1976a. Embryogenesis and formation of tetraploid and hexaploid plants from carrot protoplasts. *Can. J. Bot.* 54:1063–67.

——. 1976b. Fusion of carrot and barley protoplasts and division in heterokaryocytes. *Can. J. Genet. Cytol.* 18:263–69.

Dudits, D.; Koncz, Cs.; Bajszar, Gy.; Hadlaczky, Gy.; Lazar, G.; and Horvath, G. 1980. Intergeneric transfer of nuclear markers through fusion between dividing and mitotically inactive plant protoplasts. In *Advances in protoplast research*, ed. L. Ferenczy and G. L. Farkas, pp. 307–14. Budapest: Akademiai Kiado.

Dudits, D.; Lazar, G.; and Bajszar, G. 1979. Reversible inhibition of somatic embryo differentiation by bromodeoxyuridine in cultured cells of *Daucus carota* L. *Cell Differentiation* 8:135–44.

Dudits, D.; Rasko, I.; Hadlaczky, Gy.; and Lima-de-Faria, A. 1976. Fusion of human cells with carrot protoplasts induced by polyethylene glycol. *Hereditas* 82:121–24.

Evans, D. A.; Wetter, L. R.; and Gamborg, O. L. 1980. Somatic hybrid plants of *Nicotiana glauca* and *Nicotiana tabacum* obtained by protoplast fusion. *Physiol. Plant.* 48:225–30.

Fernandez, S. M.; Lurquin, P. F.; and Kado, C. I. 1978. Incorporation and maintenance of recombinant-DNA plasmid vehicles pBR313 and pCR1 in plant protoplasts. *FEBS Lett.* 87:277–82.

Fozard, J. R.; Part, M.; Prakash, N. J.; Grove, J.; Schechter, P. J.; Szoerdsma, A.;

and Koch-Weser, J. 1980. L-ornithine decarboxylase: an essential role in early embryogenesis. *Science* 208:505–8.

Fujimura, T., and Komamine, A. 1975. Effects of various growth regulators on the embryogenesis in a carrot cell suspension culture. *Plant Sci. Lett.* 5:359–64.

Gleba, I. I. 1978. Extranuclear inheritance investigated by somatic hybridization. In *Frontiers of plant tissue culture 1978*, ed. T. A. Thorpe, pp. 95–102. Calgary, Alberta: International Association of Plant Tissue Culture, University of Calgary.

Gleba, Y. Y., and Hoffman, F. 1980. "Arabidobrassica," a novel plant obtained by protoplast fusion. *Planta* 149:112–17.

Gosch, G., and Reinert, J. 1976. Nuclear fusion in intergeneric heterokaryocytes and subsequent mitosis of hybrid nuclei. *Naturwissenschaften* 63:534–35.

——. 1978. Cytological identification of colony formation of intergeneric somatic hybrid cells. *Protoplasma* 96:23–38.

Grambow, H. J.; Kao, K. N.; Miller, R. A.; and Gamborg, O. L. 1972. Cell division and plant development from protoplasts of carrot cell suspension cultures. *Planta* 103:348–55.

Graves, J. M.; Chew, G. K.; Cooper, D. W.; and Johnston, P. G. 1979. Marsupial-mouse cell hybrids containing fragments of the marsupial X chromosome. *Somatic Cell Genet.* 5:481–89.

Halperin, W., and Wetherell, D. R. 1964. Ontogeny of adventive embryos of wild carrot. *Science* 147:756–58.

Harms, C. T.; Potrykus, I.; and Widholm, J. M. In press. Complementation and dominant expression of amino acid analogue resistance markers in somatic hybrid clones from *Daucus carota* after protoplast fusion. *Z. Pflanzenphysiol.*

Haydu, Zs.; Lazar, G.; and Dudits, D. 1977. Increased frequency of polyethylene glycol-induced protoplast fusion by dimethyl sulfoxide. *Plant Sci. Lett.* 10:357–60.

Izhar, S., and Power, J. B. 1979. Somatic hybridization in *Petunia*: a male sterile cytoplasmic hybrid. *Plant Sci. Lett.* 14:49–55.

Jacques, S., and Sung, Z. R. 1978. Effect of halo blight toxin on pyrimidine and arginine biosynthesis in cultured cells of *Daucus carota* L. *Proceedings of the IV International Conference on Plant Pathogenic Bacteria*, Angers, France, 2:657–61.

——. 1981. Regulation of pyrimidine and arginine biosynthesis investigated by the use of phaseolotoxin and 5-fluorouracil. *Plant Physiol.* 67:287–91.

Kao, K. N. 1977. Chromosomal behaviour in somatic hybrids of soybean – *Nicotiana glauca*. *Molec. Gen. Genet.* 50:225–30.

Kao, K. N.; Constabel, F.; Michayluk, M. R.; and Gamborg, O. L. 1974. Plant protoplast fusion and growth of intergeneric hybrid cells. *Planta* 120:215–27.

Kao, K. N., and Michayluk, M. R. 1974. A method for high-frequency intergeneric fusion of plant protoplasts. *Planta* 115:355–67.

Kinnersley, A. M., and Dougall, D. K. 1980. Increase in anthocyanin yield from wild carrot cell cultures by a selection system based on cell aggregate size. *Planta* 129:200–4.

Lawyer, A. L.; Berlyn, M. B.; and Zetlich, I. 1980. Isolation and characterization of glycine hydroxamate-resistant cell lines of *Nicotiana tabacum*. *Plant Physiol.* 66:334–41.

Lazar, G.; Varga, J.; Dudits, D.; and Sung, Z. R. 1979. Analysis of cycloheximide resistant carrot cell line by somatic cell hybridization. Paper read at the Fifth International Protoplast Symposium, 9–14 July 1979, Szeged, Hungary, p. 99.

Lazar G.; Dudits D.; and Sung Z. R. In press. Expression of cycloheximide resistance in carrot somatic hybrids and their segregants. *Genetics.*

Lewin B., ed. 1980. *Gene expression 2.* 2d ed. *Eukaryotic chromosomes.* New York: Wiley-Interscience.

Lörz, H., and Potrykus, I. 1978. Investigations on the transfer of isolated nuclei into plant protoplasts. *Theor. Appl. Genet.* 53:251–56.

Lurquin, P. F. 1979. Entrapment of plasmid DNA by liposomes and their interactions with plant protoplasts. *Nucl. Acids. Res.* 6:3773–84.

——. 1980. Recent advances in the transfer of DNA to plant protoplasts. In *Plant cell cultures: results and perspectives,* ed. F. Sala, B. Parisi, R. Cella, and O. Ciferri, pp. 247–50. Amsterdam: Elsevier/North-Holland.

Lurquin, P. F., and Marton, L. 1980. DNA transfer experiments with plant and bacterial plasmids. In *Advances in protoplast research,* ed. L. Ferenczy and G. L. Farkas, pp. 389–405. Budapest: Akademiai Kiado.

McWilliam, A. A.; Smith, S. M.; and Street, H. E. 1974. The origin and development of embryoids in suspension cultures of carrot *(Daucus carota). Ann. Bot.* 38:243–50.

Maliga, P. 1978. Resistance mutants and their use in genetic manipulation. In *Frontiers of plant tissue 1978,* ed. T. A. Thorpe, pp. 381–91. Calgary, Alberta: International Association of Plant Tissue Culture, University of Calgary.

Maliga, P.; Kiss, Z. R.; Nagy, A. H.; and Lazar, G. 1978. Genetic instability in somatic hybrids of *Nicotiana tabacum* and *N. knightiana. Molec. Gen. Genet.* 163:145–51.

Maliga, P.; Lazar, G.; Joo, F.; Nagy, A. H.; and Menczel, L. 1977. Restoration of morphogenetic potential in *Nicotiana* by somatic hybridization. *Molec. Gen. Genet.* 157:291–96.

Maliga, P.; Nagy, F.; Le Thi Xuan; Kiss, Zs. R.; Menczel, L.; and Lazar, G. 1980. Protoplast fusion to study cytoplasmic traits in *Nicotiana.* In *Advances in protoplast research,* ed. L. Ferenczy and G. L. Farkas, pp. 341–48. Budapest: Akademiai Kiado.

Malmberg, R. L. 1979. Temperature-sensitive variants of *Nicotiana tabacum* isolated from somatic cell culture. *Genetics* 92:215–21.

Malmberg, R. L., and Griesbach, R. J. 1980. The isolation of mitotic and meiotic chromosomes from plant protoplasts. *Plant Sci. Lett.* 17:141–47.

Matsumoto, H.; Gregore, D.; and Reinert, J. 1975. Changes in chromatin of *Daucus carota* cells during embryogenesis. *Phytochemistry* 14:41–47.

Matthews, B. F.; Shye, S. C. H.; and Widholm, J. W. 1980. Mechanism of resistance of a selected carrot cell suspension culture to S(2-aminoethyl)-L-cysteine. *Z. Pflanzenphysiol.* 96:453–63.

Meins, F., and Lutz, J. 1980. Epigenetic changes in tobacco cell culture: studies of cytokinin habituation. In *Genetic improvement of crops — emergent techniques*, ed. I. Rubenstein, B. Gegenbach, R. Phillips, and C. Green. Minneapolis: University of Minnesota Press.

Melchers, G. 1977. The combination of somatic and conventional genetics in breeding. *Plant Res. Develop.* 5:86–110.

Melchers, G., and Labib, G. 1974. Somatic hybridisation of plants by fusion of protoplasts. I. Selection of light resistant hybrids of "haploid" light sensitive varieties of tobacco. *Molec. Gen. Genet.* 135:277–94.

Melchers, G.; Sacristan, M. D.; and Holder, A. A. 1978. Somatic hybrid plants of potato and tomato regenerated from fused protoplasts. *Carlsberg Res. Commun.* 43:203–18.

Mok, M. C.; Gabelman, W. H.; and Skoog, F. 1976. Carotenoid synthesis in carrot tissue cultures of *Daucus carota* L.. *J. Am. Soc. Hort. Sci.* 10164:442–49.

Montague, M. J.; Armstrong, T. Z.; and Jaworski, E. G. 1979. Polyamine metabolism in embryogenic cells of *Daucus carota*. II. Changes in arginine decarboxylase activity. *Plant Physiol.* 63:341–45.

Montague, M. J.; Koppenbrink, J. W.; and Jaworski, E. G. 1978. Polyamine metabolism in embryogenic cells of *Daucus carota*. I. Changes in intracellular content and rates of synthesis. *Plant Physiol.* 62:430–33.

Müller, A. J., and Grafe, R. 1978. Isolation and characterization of cell lines of *Nicotiana tabacum* lacking nitrate reductase. *Molec. Gen. Genet.* 157:285–90.

Murashige, T., and Skoog, F. 1962. A revised medium for rapid growth and bioassays with tobacco tissue cultures. *Physiol. Plant.* 15:473–97.

Nehls, R. 1978. The use of metabolic inhibitors for the selection of fusion products of higher plant protoplasts. *Molec. Gen. Genet.* 166:117–18.

Nishi, A.; Yoshida, A.; Mori, M.; and Sugano, N. 1974. Isolation of variant carrot cell lines with altered pigmentation. *Phytochemistry* 13:1653–56.

Ohyama, K.; Gamborg, O. L.; and Miller, R. A. 1972. Uptake of exogenous DNA by plant protoplasts. *Can. J. Bot.* 50:2077–80.

Ohyama, K.; Pelcher, L. A.; and Schaefer, A. 1978. DNA uptake by plant protoplasts and isolated nuclei: biochemical aspects. In *Frontiers of plant tissue culture 1978*, ed. T. A. Thorpe, pp. 75–84. Calgary, Alberta: International Association of Plant Tissue Culture, University of Calgary.

Palmer, J. E., and Widholm, J. N. 1975. Characterization of carrot and tobacco cell cultures resistant to *p*-fluorophenylalanine. *Plant Physiol.* 56:233–38.

Potrykus, I. 1973. Transplantation of chloroplasts into protoplasts of *Petunia*. *Z. Pflanzenphysiol.* 70:364–66.

Raghavan, V., ed. 1976. *Experimental embryogenesis in vascular plants*. New York: Academic Press.

Reinert, J. C. 1959. Über die Kontrolle der Morphogenese und die Induktion von Adventurembryonen an Gewebakulturen aus Karotten. *Planta* 53:318–33.

——. 1968. Factors of embryo formation in plant tissues cultivated *in vitro*. In *Les cultures de tissues de plantes*. Colloques Nationaux du Centre National de la Recherche Scientifique, Strasburgh, 5–7 May 1967. Editions du Centre National de la Recherche Scientifique, Paris, France, pp. 33–39.

Reinert, J., and Gosch, G. 1976. Continuous division of heterokaryons from *Daucus carota* and *Petunia hybrida* protoplasts. *Naturwissenschaften* 63:534.

Rodgers, A. 1979. Detection of small amount of human DNA in human-rodent hybrids. *J. Cell Sci.* 38:391–403.

Rollo, F.; Sala, F.; Cella, R.; and Parisi, B. 1980. Liposome-mediated association of vesicle lipid composition. In *Plant cell cultures: results and perspectives*, ed. F. Sala, B. Parisi, R. Cella, and O. Ciferri, pp. 237–46. Amsterdam:Elsevier/North-Holland.

Ruddle, F. 1973. Linkage analysis in man by somatic cell genetics. *Nature* 242:165–69.

Schieder, O. 1977. Hybridization experiments with protoplasts from chlorophyll-deficient mutants of some solanaceaous species. *Planta* 137:253–57.

Schieder, O. 1978. Somatic hybrids of *Datura innoxia* Mill. and *Datura discolor* Beruh. and *Datura innoxia* Mill. and *Datura stramonium* L. var. *tatula* L. *Molec. Gen. Genet.* 167:113–19.

Schieder, O., and Vasil, I. 1980. Fusion and somatic hybridization. In *Int. Rev. Cytol.* 11A, *Perspectives in plant cell and tissue culture*, ed. I. K. Vasil. New York: Academic Press.

Schilperoort, R. A.; Klapwijk, P. N.; Hooykaas, P. J. H.; Kockman, B. P.; Ooms, G.; Otten, L. A. B. M.; Wurzer-Figurelli, E. M.; Wullems, G. J.; and Rorsch, A. 1978. *A. tumefaciens* plasmids as vectors for genetic transformation of plant cells. In *Frontiers of plant tissue culture 1978*, ed. T. A. Thorpe, pp. 85–94. Calgary, Alberta: International Association of Plant Tissue Culture, University of Calgary.

Schwartz, A. G.; Coock, P. R.; and Harris, H. 1971. Correction of genetic defect in a mammalian cell. *Nature New Biol.* 230:5–8.

Sengupta, C., and Raghavan, V. 1980a. Somatic embryogenesis in carrot cell suspension. I. Pattern of protein and nucleic acid synthesis. *J. Exp. Bot.* 31:247–58.

——. 1980b. Somatic embryogenesis in carrot cell suspension. II. Synthesis of ribosomal RNA and poly(A)$^+$ RNA. *J. Exp. Bot.* 31:259–68.

Smith, J.; Furner, I.; and Sung, Z. R. In press. Nutritional and karyotypic characterization of a haploid carrot culture. *In Vitro.*

Steward, F. C.; Mapes, M. D.; and Smith, J. 1958. Growth and organized development of cultured cells. II. Organization in cultures grown from freely suspended cells. *Am. J. Bot.* 45:705–8.

Straub, J. 1971. Überweibchen aus Sortenkreuzungen bei Männlich sterilen Möhren. *Die Naturwissenschaften* 1:57–58.

Street, H. E. 1977. *Plant tissue and cell culture.* Berkeley: University of California Press.

Sung, Z. R. 1975. Auxin-independent growth in a carrot cell line resistant to DL-5-methyltryptophan. *Plant Physiol.* 56:37.

——. 1976a. Mutagenesis of cultured plant cells. *Genetics* 84:51–57.

——. 1976b. Turbidimetric determination of plant cell culture growth and its applications. *Plant Physiol.* 57:460–62.

——. 1979. Relationship of indole-3-acetic acid and tryptophan concentrations in normal and 5-methyltryptophan-resistant cell lines of wild carrots. *Planta* 145:339–45.

Sung, Z. R.; Horowitz, J.; and Smith, R. 1979. Quantitative studies in normal and 5-methyltryptophan-resistant cell lines of wild carrot: the effects of growth regulators. *Planta* 147:236–40.

Sung, Z. R., and Jacques, S. 1980. 5-Fluorouracil resistance in carrot cell culture: its use in studying the interaction of the pyrimidine and arginine pathways. *Planta* 148:389–96.

Sung, Z. R.; Lazar, G. B.; and Dudits, D. 1981. Cycloheximide resistance in carrot culture: a differentiated function. *Plant Physiol.* 68:261–64.

Sung, Z. R., and Okimoto, R. 1981. Embryonic proteins in carrot somatic embryos. *Proc. Nat. Acad. Sci. (USA)* 70:2330–35.

Suzuki, M., and Takebe, I. 1976. Uptake of single-stranded bacteriophage DNA by isolated tobacco protoplasts. *Z. Pflanzenphysiol.* 78:421–33.

Szabados, L., and Dudits, D. 1980. Fusion between interphase and mitotic plant protoplasts. Induction of premature chromosome condensation. *Exp. Cell Res.* 127:442–46.

Szabados, L.; Hadlaczky, Gy.; and Dudits, D. 1981. Uptake of isolated plant chromosome by plant protoplasts. *Planta* 151:141–45.

Thorpe, T. A. 1980. Organogenesis *in vitro*: structural, physiological and biochemical aspects. In *Int. Rev. Cytol.* 11A, *Perspectives in plant cell and tissue culture*, ed. I. K. Vasil, pp. 71–112. New York: Academic Press.

Tsumamoto, K.; Klein, R.; and Hatanka, M. 1980. Insertion of *muntjac* gene segment into hamster cells by cell fusion. *J. Cell. Physiol.* 104:225–32.

Uchimiya, H., and Murashige, T. 1977. Quantitative analysis of the fate of exogenous DNA in *Nicotiana* protoplasts. *Plant Physiol.* 59:301–8.

Vasil, I. K.; Ahuja, M. R.; and Vasil, V. 1979. Plant tissue cultures in genetics and plant breeding. *Adv. Genet.* 20:127–217.

Wallin, A.; Glimelius, K.; and Eriksson, T. 1974. The induction of aggregation and fusion of *Daucus carota* protoplasts by polyethylene glycol. *Z. Pflanzenphysiol.* 74:64–80.

Ward, M.; Davey, M. R.; Mathias, R. J.; Cocking, E. C.; Clothier, R. H.; Balls, H.; and Lucy, J. A. 1979. Effects of pH, Ca^{2+}, temperature and protease pretreatment on interkingdom fusion. *Somatic Cell Genet.* 5:529–36.

Widholm, J. M. 1972. Anthranilate synthetase from 5-methyltryptophan-susceptible and -resistant cultured *Daucus carota* cells. *Biochim. Biophys. Acta* 279:48–57.

——. 1974. Cultured carrot cell mutants: 5-methyltryptophan-resistant trait carried from cell to plant and back. *Plant Sci. Lett.* 3:323–30.

——. 1976. Selection and characterization of resistance to lysine, methionine, and proline analogs. *Can. J. Bot.* 54:1523–29.

——. 1977a. Relation between auxin-autotrophy and tryptophan accumulation in cultured plant cells. *Planta* 134:103–8.

——. 1977b. Selection and characterization of biochemical mutants. In *Plant tissue culture and its bio-technological applications*, ed. W. Barz, E. Reinhard, and M. H. Zenk, pp. 112–122. Springer-Verlag.

Willecke, K. 1978. Results and prospects of chromosomal gene transfer between cultured mammalian cells. *Theor. Appl. Genet.* 52:97–104.

Zelcer, A.; Aviv, D.; and Galun, E. 1978. Interspecific transfer of cytoplasmic male sterility by fusion between protoplasts of normal *Nicotiana sylvestris* and X-ray irradiated protoplasts of male-sterile *N. tabacum. Z. Pflanzenphysiol.* 90:397–407.

3

METHODS FOR PREPARATION OF PLANT NUCLEIC ACIDS OPTIMALLY SUITED FOR RESTRICTION ENDONUCLEASE DIGESTING AND CLONING: THE CONSTRUCTION OF JACK BEAN AND SOYBEAN PHAGE LIBRARIES IN CHARON 4A

Michael T. Sung
Jerry L. Slightom

INTRODUCTION

Until recently the direct study of eukaryotic gene organization and structure has been formidable. The size and complexity of eukaryotic genomes precludes the isolation of a specific deoxyribonucleic acid (DNA) fragment by conventional methods. However, with the development of recombinant DNA technology, studies at the molecular level of even the largest and most complex eukaryotic genomes have become feasible. Simply, molecular cloning permits the isolation of specific DNA fragments from complex genomes such as bacterial clones. Large quantities of these DNA fragments can be obtained using the replication machinery of the host bacterium.

Applications of molecular cloning have already led to some revolutionary discoveries about the organization and structure of many eukaryotic genes from animals. These discoveries include: the organization of sea urchin (Cohn, Lowry, and Kedes 1976; Schaffner et al. 1978) and *Drosophila* (Lifton et al. 1977) histone genes in tandem repeats; the presence of intervening sequences in the chicken ovalbumin gene (Breathnack, Mandel, and

This research was supported in part by the Illinois Soybean Program Operating Board.

Chambon 1977) and other genes; the organization of globin gene families in human (Fritsch, Lawn, and Maniatis 1980; Lauer, Shen, and Maniatis 1980) and rabbit (Lacy et al. 1979); rearrangement of immunoglobin gene sequences during cellular differentiation (Brack et al. 1978; Early et al. 1980; Max, Seidman, and Leder 1979); and the presence of pseudo genes (Hardison et al. 1979; Vanin et al. 1980). Thus far the study of eukaryotic genes by recombinant DNA techniques has been largely limited to the animal kingdom. Very little molecular cloning of DNA from higher plants has been reported. The large volume of research on animal genes can be attributed in part to their biomedical interest. However, the major reason for the lack of molecular cloning of higher plant DNAs appears to be due to the low cloning efficiencies found with plant DNAs (see below). Generally speaking, the cloning efficiency for plant DNA is about one to three orders of magnitude lower than that for bacterial and animal DNA, using either plasmid (Gerlach and Bedbrook 1979) or lambda vectors (see below). This lower cloning efficiency of plant DNAs has been attributed to intrinsic factors unique to plant cells and tissues that reduce the size and purity of the isolated DNA. Plant cells are much more resistant to breakage and contain large amounts of nucleases and contaminants such as oils, pigments, secondary metabolites, polyphenols, and polysaccharides that copurify with the DNA.

There are a limited number of published procedures for the isolation of plant DNA. Generally, these methods are modifications of protocols already in use for the isolation of DNA from animal tissues. The usefulness of these modified methods is limited to one or at best only a few specific plant tissues: they are not comprehensive enough to handle the overall problems associated with plant DNA isolation. Two methods widely used for the preparation of plant DNA are a modified Marmur procedure (Darby et al 1970) and a hydroxyapatite method developed by Britten, Povich, and Smith (1970). The modified Marmur technique employs cetyltrimethylammonium bromide (CTAB) to precipitate DNA from plant cell lysates. However, CTAB is also known to cleave DNA at 5-methylcytosine residues (Katterman 1975) and, therefore, the average size of the DNA is reduced to 20,000 bp. Britten's method yields shorter DNA molecules, with an average size of about 10,000 bp (Walbot and Goldberg 1978). Neither of these methods gives DNA of a size suitable for the construction of a genomic DNA library (see below).

Recently, several laboratories have shown that plant DNA suitable for restriction endonuclease digestion and cloning can be obtained from maize (Shelton and Stout, unpubl.), tobacco (Yadav et al. 1980), and sunflower tissue culture cells (Ursic, Slightom, and Kemp. unpubl.). In these cases, DNAs were isolated using conventional methods such as Marmur's (1961 procedure for maize, a pronase-Sepharose 4B procedure developed by Heyn, Hermans, and Schilperoort (1974) for tobacco (Chilton et al.

1977), and the DNA isolation procedure developed by Bendich, Anderson, and Ward (1980) for sunflower. However, when methods similar to the above were used to isolate DNAs from plant tissues themselves, the resulting DNA exhibited discontinuous kinetics of digestion with restriction endonucleases, and the cloning efficiencies were three to four orders of magnitude lower than that for control DNA (our unpublished data). These results suggest that plant DNAs suitable for cloning can be obtained without much difficulty from plant tissue culture cells, while the use of plant tissues themselves for DNA isolation is still problematical. A comprehensive method, therefore, needs to be developed for the isolation of DNA from diverse plant cells and tissues.

We describe here in detail a procedure for the isolation of DNA from jack bean embryos. The method involves breakage of cells in a dry or frozen state (effectively reducing the activity of plant nucleases), rapid deactivation of plant nucleases in a cell lysis solution, and removal of contaminating polysaccharides. Our method yields a DNA preparation that is 50 kbp in length, exhibits linear kinetics of digestion with restriction endonucleases, and has a cloning efficiency equivalent to that of animal DNAs. We then describe the necessary modifications of this procedure to enable the isolation of DNA from soybean, a plant tissue from which we and others have found DNA isolation to be extremely difficult. Again, the method yields soybean DNA suitable for constructing a genomic library (see below).

ISOLATION OF JACK BEAN DNA

The procedure as outlined in Table 3.1 is designed for 3 g dry embryos, which were individually hand dissected from jack bean seeds (purchased from Sigma Chemical Company, St. Louis). All operations can be performed at room temperature. The embryos were dry ground in a large mortar and pestle. (The grinding should be very thorough in order to increase DNA recovery from the embryos.) Freshly prepared Lysis Buffer (step 2) was added, and, upon gentle stirring, the solution became highly viscous owing to the solvation of chromatin in guanidine-HC1. After heating at 55°C for 1 hour, the viscosity was reduced, indicating the dissociation of DNA from chromosomal proteins. (If the length of heating is not adequate, DNA can be lost because chromatin will sediment above the DNA band in the cesium cushion—see step 3.) The composition of the Lysis Buffer was designed to prevent DNA degradation by plant deoxyriblonucleases (DNases) and activation of DNase by divalent metal ions and also to maximally solubilize dry plant tissues.

It is known that 7 M guanidine-HCl is a most efficient protein de-

TABLE 3.1————————————————
Jack Bean DNA Isolation Protocol

1. Dry ground 3 g jack bean embryos to a fine powder.

2. Add 40 ml of Lysis Buffer (7 M guanidine-HCl [absolute grade from Heico]; 2 percent sarkosyl; 20 mM EDTA; 20 mM Tris-HCl, pH 8.0]. Heat at 55°C for 1 hour with occasional stirring. Add 8 ml of H_2O and centrifuge for 20 minutes at 10,000 rpm in Sorvall HB4 swinging bucket rotor. Then collect and pool the supernatants.

3. Layer 7 ml of the supernatant on top of a CsCl step-gradient consisting of 2 ml of 2.85 M ($\varrho = 1.35$) CsCl, 0.02 M Tris-HCl, pH 8.0; and 2 ml of 5.7 M ($\varrho = 1.7$) CsCl, 0.02 M Tris-HCl, pH 8.0. Centrifuge in a Beckman SW 41 rotor at 38,000 rpm, 20°C for 24 hours.

4. Collect 0.4 ml fractions and electrophorese 10 μl from each fraction on a 0.7 percent agarose gel to identify the DNA-containing fractions. Pool the DNA-containing fractions and dilute them twofold with H_2O and then add an equal volume of isopropanol. Spool the DNA out on a glass rod; then dissolve the DNA in DNA dialysis buffer (0.01 M tris-HCl, pH 8.0; 0.01 M NaCl; 0.001 M EDTA).

5. Deproteinize by adding 0.5 mg/ml of Proteinase K and incubating at 55°C for 20 minutes in the presence of 0.5 percent SDS. Extract 1× with phenol (redistilled under N_2 and stored at $-20°C$ with 0.1 percent 8-hydroxy-quinoline), then 1× with phenol:chloroform:isoamyl alcohol (50:48:2), and then 1× with chloroform:isoamyl alcohol (96:4). After each extraction centrifuge the emulsion at 10,000 rpm in a Sorvall HB4 rotor for 10 minutes. Spool out DNA with a glass rod upon the addition of 1 volume of isopropanol.

6. Dissolve DNA in 4 ml DNA dialysis buffer and then add CsCl 0.95 g/ml, ethidium bromide 0.5 mg/ml, and β-mercaptoethanol 10 μm/ml. Centrifuge at 55,000 rpm, 24 hours in a Beckman Type 70.1 Ti rotor at 20°C. Recover the fluorescent DNA band by side-puncture of the centrifuge tube.

7. Remove the ethidium bromide from the DNA by extracting it with isopropanol, previously saturated with CsCl. Three extractions are usually sufficient. Dilute with 3 volumes of H_2O and add an equal volume of isopropanol to precipitate DNA. Dissolve the DNA in 0.3 M NaOAc and reprecipitate with ethanol. Dry under vacuum and then dissolve in DNA dialysis buffer at a concentration of 500 μg/ml.

Source: Compiled by the authors.

naturant; denaturant is further enhanced by the detergent "sarkosyl" (sodium lauryl sarkosinate). The Lysis Buffer should be prepared freshly because 2 percent sarkosyl is sparingly soluble in 7 M guanidine-HCl, and upon standing at room temperature sodium sarkosinate crystallizes. (Also note that sodium-dodecylsulfate [SDS] is not soluble in 7 M guanidine-HCl and therefore cannot be substituted.) We have noticed that guanidine-HCl is excellent in maintaining starch in solution, but when this 7 M guanidine-

HCl mixture is layered directly onto a gradient and centrifuged, starch precipitates out at the cesium chloride interface. This precipitation results in the loss of DNA by entrapment at this interface. We eliminated this problem by reducing the guanidine-HCl concentration to 5.8 M, and insoluble starch thus formed was removed by low-speed centrifugation (step 2). The solution volumes specified in Table 3.1 for sedimenting DNA into a CsCl cushion will fill all six buckets of the Beckman SW 41 rotor. This swinging bucket rotor is most effective owing to its relatively large capacity and the centrifugal force needed to band DNA. A 2-ml intermediate density CsCl was used (step 3) to prevent the mixing of contaminants from the guanidine-sarkosyl cell lysate layer with banded DNA in the lower cesium cushion.

In many attempts to purify soybean DNA (see below) we have verified that the method of sedimenting DNA through a clean CsCl cushion and banding it in the lower CsCl cushion is superior to the method of mixing cell and tissue lysate with CsCl and then banding the DNA. The DNA recovered after alcohol precipitation (step 4) had an absorbance ratio A_{260}/A_{280} of 1.58. The residual protein contaminant was removed by proteinase K treatment followed by phenol, chloroform, and isoamyl alcohol extractions (step 5), and the ratio was then increased to 2.0. Also note that proteinase K digestion was done at high enzyme concentration (0.5 mg/ml) in the presence of 0.5 percent SDS and high temperature to inhibit DNase. The DNA was spooled onto a glass rod upon the addition of an equal volume of isopropanol, which effectively removed any low molecular weight DNA and/or ribonucleic acid (RNA). Polysaccharides that may have cobanded with the DNA at $\varrho = 1.7$ in step 3 were removed by an equilibrium centrifugation in the presence of ethidium bromide (step 6). The intercalation of ethidium bromide lowers the buoyant density of the DNA to $\varrho = 1.45$. The DNA yield was 100 μg per gram embryos.

ISOLATION OF SOYBEAN DNA

When identical procedures were used for the isolation of DNA from soybean embryos, the product was noticeably impure and the cloning efficiency was three orders of magnitude lower than that for jack bean DNA (see below). It appears that soybean embryos contain either more or different contaminants than jack bean embryos, which inhibit cloning. In the following we describe modifications made to the above jack bean protocol for the removal of extra contaminants found in soybean embryos (see Table 3.2).

Soybean embryos contain a large amount of oil, pigments, and polysaccharides including starch, pectin, and galactomannan. A major portion of oil, pigment, and polysaccharides are removed by the guanidine-sarkosyl

TABLE 3.2
Soybean DNA Isolation Protocol

1. Dry ground 2 g soybean embryos to a fine powder.

2. Add 40 ml of Lysis Buffer (7 M guanidine-HCl; 2 percent sarkosyl; 20 mM EDTA; 20 mM Tris-HCl, pH 8.0) and incubate at 55°C for 1 hour. Then add 8 ml H_2O and centrifuge for 20 minutes at 10,000 rpm in Sorvall HB4 swinging bucket rotor.

3. Add ethidium bromide to 0.5 mg/ml and β-mercaptoethanol to 10 mM, mix, and then layer 7 ml of this mixture on top of 6 CsCl step-gradients. Each gradient consists of 3 ml of 2.85 M ($\varrho = 1.35$) CsCl, 0.02 M Tris-HCl (pH 8.0) on top of 1 ml of 5.7 M ($\varrho = 1.7$) CsCl, 0.02 M Tris-HCl (pH 8.0). Centrifuge in a Beckman SW 41 rotor at 38,000 rpm, 20°C for 24 hours. Aspirate off the top 7 ml and very gently withdraw the DNA band with a blunt-end 16-gauge needle.

4. Add 5 g of guanidine-HCl and adjust sarkosyl to 1 percent in a final volume of 10 ml. Heat the mixture at 55°C for 0.5 hour and then add ethidium bromide to 0.5 mg/ml and band the DNA ($\varrho = 1.5$) in a Beckman Type 70.1 Ti rotor at 55,000 rpm, 20°C for 24 hours. Remove DNA band by side-puncture and extract ethidium bromide 3× with isopropanol saturated with CsCl. Dilute the DNA solution with 3 volumes of H_2O and precipitate DNA with equal volumes of isopropanol. Dissolve the DNA in DNA dialysis buffer.

5. Deproteinize by incubating with 0.5 mg/ml of proteinase K at 55°C for 20 minutes in the presence of 0.5 percent SDS. Extract 1× with phenol, then 1× with phenol:chloroform:isoamyl alcohol (50:48:2), and then 1× with chloroform: isoamyl alcohol (96:4). After each extraction centrifuge the emulsion at 10,000 rpm in a Sorvall HB4 rotor for 10 minutes. Spool out DNA with glass rod upon addition of 1 volume of isopropanol.

6. Dissolve DNA in 4 ml DNA dialysis buffer and then add CsCl to $\varrho = 1.7$. Centrifuge at 30,000 rpm, 20°C for 48 hours in Beckman Type 70.1 rotor. After centrifugation collect 0.3 ml fractions by puncturing the bottom of the centrifuge tube. Locate the DNA-containing fractions by agarose gel electrophoresis.

7. Pool and dilute the DNA-containing fractions with 3 volumes of H_2O. Precipitate DNA with 1 volume of isopropanol. Dissolve DNA in 0.3 M NaOAc and then ethanol precipitate. Dry under vacuum and then dissolve DNA in DNA dialysis buffer at a final concentration of 500 µg/ml.

Source: Compiled by the authors.

procedure. Following ultracentrifugation (step 3 of Tables 3.1 and 3.2) oil could be seen on top of the gradient while polysaccharides and pigments were found in the guanidine-sarkosyl mixture. There was also a high-buoyant-density polymer (compound X) that sedimented into the CsCl cushion of $\varrho = 1.7$ (see Fig. 3.1). This polymer behaved like DNA; it formed a highly viscous solution and bound ethidium bromide, and in the presence of alcohol it even spooled onto a glass rod. In early experiments using the procedure in Table 3.1, we were unable to remove this material from DNA with repeated equilibrium banding in neutral and ethidium bromide CsCl gradients. Therefore, we concluded that this contaminant must be removed early in the procedure, and this is reflected in the modifications made in the soybean DNA purification protocol (see Table 3.2).

We found that compound X could be separated from DNA by the addition of ethidium bromide, which shifted the buoyant density of DNA to $\varrho = 1.45$. During centrifugation compound X sedimented ahead of the DNA and pelleted at the bottom of the centrifuge tube (see Fig. 1.1), while the DNA did not enter the lower CsCl cushion. Thus, DNA was effectively separated from compound X. Fig. 3.1 also shows an additional polymer contaminant that is streaming above and below the DNA band. Care was taken to very slowly withdraw the DNA band into a 1 ml syringe via a blunt-end 16-gauge needle and at the same time to try to limit the entrance of the contaminant into the syringe and needle.

A second cycle of heating with 5 M guanidine-HCl and 1 percent sarkosyl and rebanding in ethidium bromide-CsCl density gradients (step 4, Table 3.2) was performed to further remove contaminants that had been carried over from step 3. We included a last CsCl ($\varrho = 1.7$) equilibrium centrifugation (step 6) to remove contaminants that banded with the DNA in the ethidium bromide-CsCl gradient at a $\varrho = 1.45$. We were able to isolate 200 μg of DNA per gram embryos.

METHODS OF MOLECULAR CLONING

Two approaches for cloning a specific fragment of DNA from a complex genome have been generally used. The first involves the partial purification of the DNA fragment of interest, while the second requires the cloning of the complete genome (construction of a library of clones) from which the DNA fragment of interest can be isolated. Both methods rely on DNA or RNA/DNA hybridization of a complementary radioactive DNA or RNA probe to detect the DNA fragment of interest. Radioactive probes can be obtained by the synthesis of complementary DNA (cDNA) from messenger RNA (mRNA) with reverse transcriptase (Kacian and Myers 1976) or by nick-translation of cloned cDNAs (Maniatis, Jeffrey, and Kleid 1975; and see Slightom, Blechl, and Smithies 1980).

Partial enrichment (about 100- to 1,000-fold) of DNA fragments can be accomplished by using agarose gel electrophoresis (Tonegawa et al. 1977) and RPC-5 column chromatography (Leder et al. 1977). This method requires fewer clones to be constructed and screened to obtain the DNA frag-

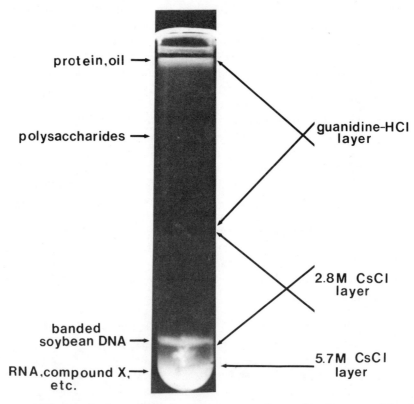

protein,oil ⟶

polysaccharides ⟶

guanidine-HCl
layer

2.8M CsCl
layer

banded
soybean DNA ⟶

5.7M CsCl
layer

RNA,compound X, ⟶
etc.

FIG. 3.1. Soybean DNA banding in guanidine/sarkosyl cesium chloride-
ethidium bromide gradient. DNA was released from dry ground soybean embryo by
heating at 55°C in Lysis Buffer (for details see Table 3.2). The guanidine/sarkosyl extract was mixed with ethidium bromide and layered on top of two CsCl steps: 3 ml of
ϱ = 1.35 upper CsCl and 1 ml of ϱ = 1.7 lower CsCl cushions in a 11.5 ml polyal-
lomer centrifuge tube. Centrifugation was for 24 hours, 35,000 rpm, at 20°C in a
Beckman SW 41 rotor. Following centrifugation the tube was illuminated with a long
wavelength ultraviolet source and photographed. The highly fluorescent DNA is
shown banding above the ϱ = 1.7 CsCl cushion. "Compound X" and RNA form a
pellet at the bottom of the centrifuge tube; the pellet is highly fluorescent. Oil and
protein can be seen on top of the gradient. DNA was removed from the gradient as
described in Table 3.2 and the text.

ment of interest. However, the genomic DNA must be fragmented (usually
by restriction endonuclease digestion) before it can be fractionated. If the
enzyme selected cuts the gene of interest, two or more DNA fragments must
be cloned to obtain a complete gene. This can be even more complicated if
the gene is repeated. Consequently, this method cannot be used to isolate
overlapping clones linking a family of genes.

The concept of cloning the entire genome eliminates the above problem and allows the full potential of molecular cloning techniques to be used. The recent isolation of clones containing overlapping DNA fragments that show the linkage relationships of seven human β globin genes (Fritsch, Lawn, and Maniatis 1980), four rabbit β globin genes (Lacy et al. 1979), and four human α globin genes (Lauer, Shen, and Maniatis 1980) demonstrates the utility of the method. Clearly, the construction of genomic libraries is the method of choice in gene cloning.

The strategy for constructing genomic libraries requires isolating long DNA molecules (50 kbp or longer), fragmenting the DNA randomly by either mechanical shearing or by nonlimit restriction endonuclease digestion, and incorporating these fragments into an *Escherichia coli* host-vector system. The number of clones needed to represent a complete genome depends on the size of the genome and the size of the random fragments to be cloned. Using the equation derived by Clarke and Carbon (1976)

$$N = \frac{\ln (1 - P)}{\ln (1 - f)}$$

the number of clones (N) needed can be calculated for a set probability (P) by knowing each fragment represents a fraction (f) of the total genome. The number of clones needed for complete libraries of various genomes are given in Table 3.3. As the genome sizes increase, the number of clones in-

TABLE 3.3
Number (N) of Library Clones Needed to Represent Various Complete Genomes at a Probability Level of .99

DNA Source	Genome Size in Base Pair	Reference*	Fragment Size (f) in Base Pairs	N (at .99)
E. coli	4.5×10^6	a	18,000†	1,150
Yeast	1.0×10^7	b	18,000	4,860
Drosophila	1.65×10^8	c	18,000	42,000
Pea	3.9×10^8	d	18,000	100,000
Soybean	1.8×10^9	d	18,000	460,000
French bean	1.9×10^9	f	18,000	490,000
Mammals	3.0×10^9	e	18,000	770,000
Wheat	5.1×10^9	d	18,000	1,300,000
Rye	6.8×10^9	d	18,000	1,720,000
Broad bean	4.4×10^{10}	d	18,000	11,000,000

*a. Cairns (1963); b. Kaback and Halvorson (1977); c. Maniatis et al. (1978); d. Walbot and Goldberg (1978); e. Watson (1970); f. Sung, Slightom, and Hall (1980).

†This number was selected because it is the average size cloned into Charon 4A. (See discussion below.)

Source: Compiled by the authors.

creases rapidly. This creates two new problems: how to generate a collection of 1×10^6 or more independent recombinant DNA clones and how to screen such a large collection of clones.

The construction of genomic libraries containing 1×10^6 clones using plasmid cloning systems (Clarke and Carbon 1976) is, from a practical viewpoint, impossible. The efficiency of plasmid transformation is relatively poor, and, therefore, a large number of competent bacteria cells are required in order to obtain a library of clones (see Blattner et al. [1978] for a discussion of this problem). Also the slow rate at which plasmids can be screened, even using the relatively rapid method developed by Grunstein and Hogness (1975), prevents their use in the construction of genomic libraries.

Both the cloning efficiency and screening problems were effectively solved by using *E. coli* phage lambda as a cloning vector. The efficiency for transducing in vitro packaged lambda recombinant DNA molecules into *E. coli* (Becker and Gold 1975; Blattner et al. 1978; Sternberg, Tiemeier, and Enquist 1977) is about three orders of magnitude higher than plasmid transformation. Furthermore, Benton and Davis (1977) greatly simplified the screening of genomic libraries when they found that DNA from a lambda plaque could be transferred onto a nitrocellulose filter and that its location could be detected by DNA or RNA/DNA hybridization. Blattner et al. (1978) scaled up this technique using cafeteria trays (43 cm × 33 cm) in place of petri dishes, so that 10^6 clones (usually more than one complete genome, see Table 3.3) could be effectively prepared for screening in a matter of a few minutes.

Numerous lambda cloning vectors have been developed recently, with the most widely used being the Charon series (Blattner et al. 1977; DeWit et al. 1980). Of this series, Charon 4A is the most useful for constructing genomic libraries because of its large capacity for passenger DNA, which is 8 to 22 kbp (Blattner et al. 1977). We describe below a general procedure for the construction of genomic libraries using Charon 4A as the cloning vector.

CONSTRUCTION OF GENOMIC LIBRARIES

Before a library can be constructed, the genomic DNA must be cut into fragments acceptable to the cloning vector. The simplest method for generating fragments for cloning into Charon 4A is to perform a nonlimit *Eco*RI digest of the DNA, followed by isolation of fragments in the range of 15 to 22 kbp. There are two problems associated with this method: some *Eco*RI sites may be preferentially cleaved and some *Eco*RI fragments representing only a few percent of the total genomic DNA are larger than the capacity of

Charon 4A even after complete digestion of the DNA. The first problem can be reduced by randomizing the EcoRI digest (see below). The second problem can be avoided by obtaining a size estimate of the EcoRI fragment that contains the DNA region of interest—for example a genomic blot analysis (Southern 1975). If the DNA fragment of interest is too large to be cloned into Charon 4A, other Charon cloning vectors accepting DNA fragments resulting from other restriction endonuclease digestions can also be used (see DeWit et al. 1980). Moreover, Blattner and coworkers have constructed a new Charon vector, number 30, which is ideal for the construction of genomic libraries. This vector will accept DNA fragments generated from an MboI digest (Frederick Blattner 1980: personal communication). Charon 30 was not available when we constructed the libraries discussed here, but the methods described below for Charon 4A also apply to Charon 30.

Before Charon 4A can be used as a cloning vector, the nonessential Bio (7.8 kbp) and Lac (6.9 kbp) EcoRI fragments must be removed, which leaves room for 22 kbp of passenger DNA (Blattner et al. 1977). To prevent recloning the Bio and Lac fragments, the Charon 4A "arms" (annealed left and right ends) are purified on a linear 5 to 20 percent NaCl gradient. An outline for the purification of annealed Charon arms from internal fragments is presented in Table 3.4. NaCl was used to form the gradients because it has a higher capacity (accommodates three to four times more DNA than sucrose gradients) to fractionate DNA without smearing (Blattner 1968). Fig. 3.2 shows the clear separation of the annealed Charon 4A arms from the Bio and Lac fragments; at this stage the arms are about 90 percent pure. After a second salt gradient centrifugation Charon 4A arms are essentially free of contamination from the Bio and Lac fragments.

Genomic DNA fragments were prepared by a procedure similar to that described above for Charon 4A arms, except that the EcoRI digestion was randomized by performing a series of nonlimit digestions and only one NaCl gradient was used. The series of partial EcoRI digestions requires about 300 μg of genomic DNA, although as little as 100 μg can be used. Each sample in this series contains about 40 μg of genomic DNA in EcoRI salts (90 mM Tris-HCl, pH 8.0; 10 mM MgCl$_2$; and 10 mM NaCl) and is digested to 5, 20, 30, 50, and 80 percent of completion by adding a calibrated amount of EcoRI enzyme and incubating at 37°C for a corresponding length of time. In addition, one sample of genomic DNA, about 80 μg, is digested to completion so that the library will contain the larger EcoRI fragments. Digestions were stopped quickly by adding diethylpyrocarbonate and EDTA to a final concentration of 0.1 percent and 10 mM, respectively, followed by heating to 68°C for 10 minutes. The percentage of completion of each digest can be estimated by visual inspection of the DNA following agarose gel electrophoresis. Fig. 3.3 illustrates gel electrophoresis to measure the kinetics for EcoRI digestion of soybean DNA isolated by the above pro-

TABLE 3.4——————————
Preparation of Charon 4A Arms

A. *Eco*RI digestion of Charon 4A phage DNA.
1. Completely digest Charon DNA with *Eco*RI. Check for complete digestion on an agarose gel.
2. If digestion is complete, deactivate *Eco*RI enzyme by adding diethyl-pyrocarbonate to a final concentration of 0.1 percent and heating to 37°C for 10 minutes.

B. Separation of Charon arms from Bio and Lac fragments.
1. Make 5 and 20 percent NaCl solutions in DNA dialysis buffer (0.01 M Tris pH, 8.0; 0.001 M EDTA).
2. From a linear gradient maker, pump 5 ml of 5 percent NaCl and 5 ml of 20 percent NaCl for each gradient. Note: These volumes are for use with the Beckman SW 41 rotor, which holds a total of 11 ml. Form the gradient by pumping the less dense NaCl solution into the bottom of the tube first, followed by the more dense NaCl solution. The less dense solution will rise and layer on top of the more dense solution as it is pumped in.
3. Load digested Charon 4A DNA onto gradients. The volume should not exceed 300 μl, and the amount of DNA should not exceed 300 μg per gradient tube. Spin at 35K, 20°C for 6 to 6.5 hours.
4. After the centrifugation collect twenty 0.5 ml fractions from each gradient tube and load 5 to 10 μl of these fractions directly onto a 0.7 percent agarose gel and electrophorese.
5. A profile similar to that shown in Fig. 3.2 should be obtained, with the Charon 4A arms near the bottom of the gradient and the Bio and Lac fragments near the top.
6. Pool fractions containing the major portion of Charon arms and measure the volume. Then add an equal volume of sterile double distilled water and mix; then add 2.5 volumes of 100 percent ethanol.
7. Store this solution at −70°C for at least 2 hours (overnight is better); then centrifuge in a SS-34 rotor at 10K, −10°C for 2 hours. Pour off supernatant.
8. Resuspend DNA pellet in 0.5 ml of 0.3 M NaOAc and reprecipitate in a 1.5 ml Eppendorf tube. Chill sample at −70°C for 10 minutes prior to spinning in the Eppendorf centrifuge. Pour off supernatant and dry DNA under vacuum.
9. Resuspend DNA pellet in DNA dialysis buffer and adjust the volume so that the DNA concentration is about 1 mg/ml.
10. Repeat steps 2 through 9. After completion of this second gradient fractionation, the Charon 4A arms should contain less than 1 percent of the original Bio and Lac fragments.

Source: Compiled by the authors.

cedure. We found the digestion kinetics for both soybean and jack bean DNAs to be comparable with that found for human DNA, using the same batch of calibrated *Eco*RI enzyme.

All *Eco*RI digestion products were combined and layered onto a 5 to

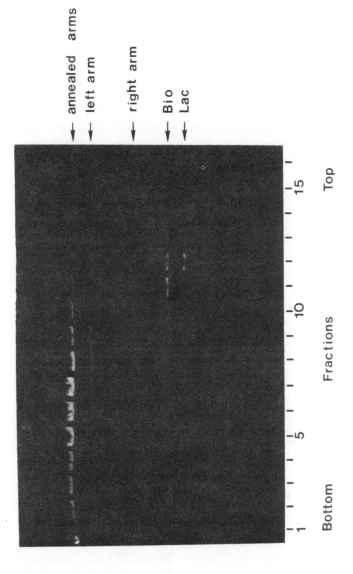

FIG. 3.2. Separation of annealed Charon 4A arms from internal Bio and Lac fragments using a 5 to 20 percent NaCl gradient. 300 μg of EcoRI digested Charon 4A DNA was layered onto a 5 to 20 percent NaCl gradient and centrifuged as described in Table 3.4. The gradient was fractionated into 0.5 ml fractions from which 5 μl was directly loaded onto a 0.7 percent agarose gel. After electrophoresis and ethidium bromide staining, fractions containing the majority of Charon 4A arms (bottom of gradient) could be distinguished from those containing the internal Bio and Lac fragments (center of gradient). Fractions containing Charon 4A arms were pooled and treated as described in Table 3.4.

20 percent NaCl gradient (not to exceed 300 μg per gradient). Conditions for centrifugation and gradient fraction collection are outlined in Table 3.4 (steps 1 to 4 in part B). Fig. 3.4 illustrates the resolution of soybean DNA fragments on a salt gradient as analyzed by loading 5 μl of each 0.5 ml fraction directly onto a 0.7 percent agarose gel. The fractions containing EcoRI genomic DNA fragments in the range of 15 to 22 kbp (as judged from the DNA size standards, see Fig. 3.4) were pooled, diluted with an equal volume of distilled H_2O, and precipitated with ethanol at $-70°C$ overnight. Genomic DNA fragments were pelleted by centrifugation in a SS-34 rotor at 10,000 rpm, $-10°C$, for 2 hours. This DNA pellet was resuspended in 400 μl of 0.3 M sodium acetate and again precipitated with ethanol by leaving it at $-70°C$ for 1 hour before centrifuging for 10 minutes in an Eppendorf microfuge. Salt was removed by rinsing the DNA pellet with 1 ml of cold ethanol; then the pellet was dried under vacuum. DNA fragments were dissolved in sterile distilled H_2O at a concentration of 0.5 to 1 mg/ml, and we estimated that about 20 percent of the total DNA loaded on the gradient was recovered.

Ligation reaction contained 10 μg each of Charon 4A arms and 15 to 22 kbp genomic DNA fragments at a final DNA concentration of 250 to 500 μg/ml. In the reaction the ratio of Charon 4A EcoRI ends to genomic EcoRI ends was about 1:2, because the annealed Charon 4A arms are about twice (31 kbp) as large as the average genomic fragment (18 kbp). Before ligase was added, the sample was warmed to 37°C for 10 minutes to ensure that the EcoRI "sticky ends" were free. Then 5 μl of 10X ligase salts (800 mM Tris-HCl, pH 8.0; 200 mM $MgCl_2$; 150 mM DTT; and 10 mM ATP) and 1 unit of T_4 DNA ligase (New England Biolabs) were added, and the volume was adjusted to 50 μl. The reaction mixture was incubated at 4°C for 18 hours. Electrophoretic comparison of DNA from the ligation reactions with nonligated sample generally showed that about 95 percent of the DNA was ligated.

Ligated Charon 4A recombinant DNA molecules were transfected into E. coli, using the highly efficient in vitro packaging method described by Blattner et al. (1978). In this method, lambda heads and tails are obtained from lysogenic strain NS 428 (Sternberg, Tiemeier, and Enquist 1977), which is defective in "A" protein production. Without A protein, endogenous lambda DNA cannot be packaged during induction of the lysogens (Wang and Kaiser 1973), which eliminates the need for physically inactivating (by ultraviolet irradiation of) prophage DNA in the extracts. Since the cells and prophage are recombination deficient, recombinations between the Charon vector and the endogenous lambda DNA in the extracts are prevented. However, before Charon recombinant DNA molecules can be packaged, A protein must be added. We obtained A protein from the strain λdg805 (Blattner et al. 1978).

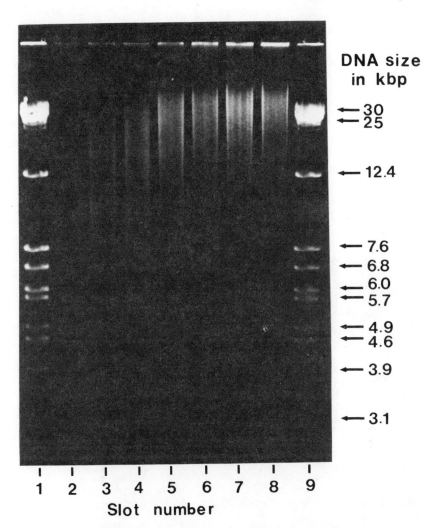

FIG. 3.3. Random partial EcoRI digestion of soybean DNA. Five samples each containing 40 μg of soybean DNA were digested with different amounts of EcoRI enzyme for different lengths of time. A sixth sample containing 80 μg of soybean DNA was digested to completion. The degree of digestion reached in each sample was analyzed by loading 1 μg of DNA onto an agarose gel. Slot numbers 2 through 7 contain EcoRI digestions estimated to be 100, 80, 50, 30 to 40, 10 to 20, and 5 percent of completion, respectively. Slot 8 contains undigested soybean DNA and slots 1 and 9 contain DNA size standards of known length (Blattner et al. 1978). Note that the undigested soybean DNA runs slower than 31 kbp; we estimate its average size to be between 40 and 50 kbp.

DNA size
in kbp

—— 25.0

—— 12.4

—— 7.6
—— 6.0

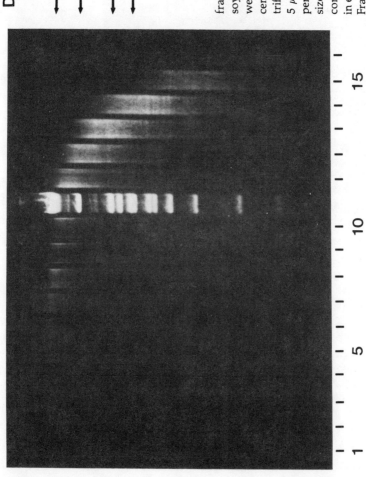

FIG. 3.4. Analysis of NaCl gradient fractionation of random partial EcoRI digested soybean DNA. Digestions shown in Fig. 3.3 were pooled and fractionated on a 5 to 20 percent NaCl gradient (see Table 3.4). After centrifugation 0.5 ml fractions were collected and 5 μl from each was loaded directly onto a 0.7 percent agarose gel. EcoRI DNA fragments the size range of 15 to 22 kbp were estimated by comparing the ethidium bromide stain pattern in each slot with DNA size standards in slot 11. Fractions represented in slots 7 through 10 were pooled and ethanol precipitated as described in the text.

15   Top

10

5

1

Bottom   Fractions

Recently two functions have been assigned to the A protein fraction: cleavage of the covalently closed lambda *cos* site and association of lambda DNA with the protein head (Becker et al. 1979). This agrees with our observation in that when nonligated Charon 4A DNA is used as a substrate for packaging, the efficiency is about 10 times higher than when ligated Charon 4A is used. Clearly, the mechanism of packaging nonligated lambda DNA is different; for this reason we suggest that ligated Charon DNA be used to test the efficiency of the in vitro packaging system. Generally, we obtain titers between 2×10^6 to 2×10^7 plaque forming units (PFU) per μg of ligated Charon 4A DNA. The efficiency of in vitro packaging or cloning efficiency can be calculated by dividing the total number of phages obtained by the number of lambda DNA molecules added to a reaction (2×10^{10} lambda molecules/μg). Thus our efficiency for packaging ligated Charon 4A DNA was between 10^{-3} to 10^{-4} (see Table 3.5).

In Table 3.5 we present data for the construction of human, mouse, chicken, jack bean, and soybean Charon 4A recombinant DNA libraries. This table contains data on the amounts of in vitro packaging components added, yields of viable phage, and the cloning efficiency for each of these libraries. From the Charon 4A-soybean in vitro packaging reaction, 5×10^5 viable recombinant phage particles were obtained, more than a complete genome representation (see Table 3.3). In the case of jack bean, the genome size is unknown, but having obtained 7×10^6 viable recombinant phage particles, we believe that a major portion, if not all, of the jack bean genome is represented in this library.

The average cloning efficiency (excluding soybean) for the construction of these genomic libraries was about 1.5×10^{-5}, which was about one-tenth that obtained for ligated Charon 4A DNA. This drop in efficiency is not surprising and is most likely due to the ligation of two Charon 4A molecules to each other at their *Eco*RI sites or to the ligation of two or more genomic *Eco*RI fragments into Charon 4A, yielding a recombinant DNA molecule too large to be packaged (larger than 55 kbp). There is no easy explanation for the 10-fold reduction in cloning efficiency for soybean DNA (1.9×10^{-6}). However, it is 100-fold better than that obtained using soybean isolated by other methods (2×10^{-8}). Jack bean DNA prepared by a less stringent procedure has the packaging efficiency of animal DNA. Thus, we can correlate the lower cloning efficiency with the degree of difficulty we have encountered in purifying the DNAs. Soybean tissue must contain either a greater concentration of contaminant or a more potent contaminant that inhibits cloning.

Potential contaminants from plant tissues are many in number, and not all can be obtained in pure form. We tested the effects of two commercially available polysaccharides (apple pectin and potato starch) on packaging ligated Charon 4A DNA. The results of these experiments are shown in

TABLE 3.5
In Vitro Packaging of Charon 4A and Recombinant Charon 4A DNA

DNA Packaged	DNA (μg) Ligated Charon 4A	Genomic	Buffer A[a]	Buffer MI[b]	Sonic[c]	pA[d]	FTL[e]	Total Viable Phage[f]	λ DNA Equivalents	Cloning Efficiency
Ligated Charon 4A	1 μg	—	30 μl	4 μl	20 μl	1 μl	150 μl	2 × 10⁶ to 20 × 10⁶	2 × 10¹⁰	1 × 10⁻⁴ to 1 × 10⁻³
Charon 4A } Human #1	10 μg	10 μg	600 μl	80 μl	400 μl	20 μl	1.50 ml	2.8 × 10⁶	2.6 × 10¹¹·⁸	1 × 10⁻⁵
Charon 4A } Human #2	10 μg	10 μg	600 μl	80 μl	400 μl	20 μl	3.0 ml	2.6 × 10⁶	2.6 × 10¹¹	1 × 10⁻⁵
Charon 4A } Mouse	10 μg	10 μg	600 μl	80 μl	400 μl	20 μl	3.0 ml	7 × 10⁶	2.6 × 10¹¹	2.7 × 10⁻⁵
Charon 4A } Chicken	10 μg	10 μg	600 μl	80 μl	400 μl	20 μl	3.0 ml	5 × 10⁶	2.6 × 10¹¹	1.9 × 10⁻⁵
Charon 4A } Jack bean	20 μg	20 μg	1.2 ml	160 μl	800 μl	40 μl	6.0 ml	6 × 10⁶	5.3 × 10¹¹	1.1 × 10⁻⁵
Charon 4A } Soybean	10 μg	10 μg	600 μl	80 μl	400 μl	20 μl	3.0 ml	5 × 10⁵	2.6 × 10¹¹	1.9 × 10⁻⁶

[a] Buffer A contains: 20 mM Tris-HCl, pH 8.0; 3 mM $MgCl_2$; 1 mM EDTA; and 0.05 percent 2-mercaptoethanol.

[b] Buffer MI contains: 0.03 M spermidine; 0.06 M putrescine; 0.018 M $MgCl_2$; 0.015 M ATP; and 0.028 M 2-mercaptoethanol. Buffer MI was adjusted to pH 7.4 by adding 2 M Tris-HCl, pH 9.0.

[c] Sonic extract containing lambda phage heads was obtained from strain NS-428; see Blattner et al. (1978).

[d] Protein A was isolated from strain λdg805; see Blattner et al. (1978).

[e] FTL (Freeze-Thaw Lysate) extract containing lambda phage tails was obtained from strain NS-428; see Blattner et al. (1978).

[f] In vitro packaging reactions were titered on E. coli strain Dp50supF.

[g] λ DNA equivalents were calculated using 31 kbp, the size of Charon 4A arms.

Note: Packaging components were added in the following order: buffer A, buffer MI, DNA, protein A; sonication and incubation at room temperature for 45 minutes; addition of FTL and incubation for at least 1 hour before titering.

Source: Compiled by the authors.

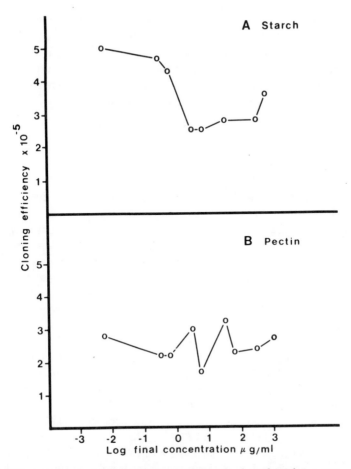

FIG. 3.5. Analysis of the effect of potato starch and apple pectin on in vitro packaging. Potato starch (Baker Analyzed reagent lot #26161) and apple pectin (BDH 250 grade) were dissolved in buffer A at a concentration of 1 mg/ml. Both potato starch and apple pectin, either from the original stock solution or dilutions of these stocks in buffer A, were added to the in vitro packaging reaction just after the addition of 1 μg of ligated Charon 4A. These samples were incubated at room temperature for 10 minutes and then the remaining packaging components were added following the procedure set in Table 3.5. Reactions were titered at three different concentrations on E. coli strain Dp50supF. Efficiencies of packaging were calculated from the average titer for each sample. The average efficiency of the control, 2 × 10^{-5}, is lower than that reported in Table 3.5. This was probably caused by adding an extra 100 μl of buffer A to each reaction. This was necessary to accommodate fluctuations in the reaction volumes caused by the addition of different amounts of starch and pectin. We have plotted the cloning efficiency found in each reaction versus the log of the final concentration (μg/ml) of either starch (**A**) or pectin (**B**) in the reactions.

Fig. 3.5 (**A** for potato starch and **B** for apple pectin). At concentrations as high as 600 μg/ml (122 times the DNA concentration) neither starch nor pectin had any effect on packaging. Thus starch and pectin are polysaccharides that do not inhibit packaging; however, plants also contain hemicellulose and other structural and storage polysaccharides. We were surprised to find that at low concentration (below 0.6 μg/ml) starch has the effect of increasing packaging efficiency (see Fig. 3.5).

CONCLUSION

In conclusion we have developed a plant DNA isolation procedure that is effective in obtaining DNA suitable for restriction endonuclease digestion and cloning from jack bean and soybean embryos. Using these DNAs we constructed partial EcoRI bacteriophage libraries in Charon 4A consisting of 7×10^6 jack bean and 5×10^5 soybean clones. These experiments, together with the work of others using DNA obtained from plant tissue culture cells (see above), clearly demonstrate that plant DNA can be cloned. The cloning efficiency has been correlated with the degree of purity in the isolated DNA.

Soybean embryo tissue presents more problems than any other plant tissue we have encountered, and the cloning efficiency of DNA isolated by conventional procedure is very poor (2×10^{-8}). By improving the DNA isolation procedure, the cloning efficiency also shows a parallel improvement, in this case a 100-fold increase (1.9×10^{-6}).

We believe the soybean DNA isolation protocol (Table 3.2) is the method of choice for the isolation of DNA from diverse plant cells and tissues. The method in principle is simple and can be adapted to wet tissues and cells by grinding in liquid N_2 and/or after lyophilization of the frozen sample. The CsCl cushions may be modified to accommodate the origin and nature of the contaminating material in a manner as we have described for removing compound X from soybean DNA. In addition, the guanidine-HCl/sarkosyl treatment can be repeated to remove even the most tenacious contaminant. Therefore, the method in its totality is comprehensive and well suited for plant materials from diverse sources.

REFERENCES

Becker, A., and Gold, M. 1975. Isolation of the bacteriophage lambda A-gene protein. *Proc. Nat. Acad. Sci. (USA)* 72:581–85.

Becker, A.; Benchimol, S.; Gold, M.; Murialdo, H.; and Sumner-Smith, M. 1979. Biochemical Studies on the Packaging and Maturation of Phage Lambda

DNA. XIth International Congress of Biochemistry. 8–13 July 1979, Toronto, Canada. (Abstract)

Bendich, A. J.; Anderson, R. S.; and Ward, B. L. 1980. Plant DNA: long, pure and simple. In *Genome organization and expression in plants*, ed. J. C. Leaver, pp. 31–33. New York: Plenum Press.

Benton, W., and Davis, R. 1977. Screening λgt recombinant clones by hybridization to single plaques in situ. *Science* 196:180–82.

Blattner, F. R. 1968. Interaction of RNA-polymerase with T_7 DNA molecules. Ph.D. dissertation, Johns Hopkins University, Baltimore, Maryland.

Blattner, F. R.; Blechl, A. E.; Denniston-Thompson, K.; Faber, H. E.; Richards, J. E.; Slightom, J. L.; Tucker, P. W.; and Smithies, O. 1978. Cloning human fetal λ globin and mouse α-type globin DNA: preparation and screening of shotgun collections. *Science* 202:1279–84.

Blattner, F. R.; Williams, B. G.; Blechl, A. E.; Denniston-Thompson, K.; Faber, H. E.; Furlong, L. A.; Grunwald, D. J.; Keefer, D. O.; Moore, D. D.; Schumm, J. W.; Sheldon, E. L.; and Smithies, O. 1977. Charon phages: safer derivatives of bacteriophage lambda for DNA cloning. *Science* 196:161–69.

Brack, C.; Hirama, M.; Lenhard-Schuller, R.; and Tonegawa, S. 1978. A complete immunoglobulin gene is created by somatic recombination. *Cell* 15:1–14.

Breathnack, R.; Mandel, J. L.; and Chambon, P. 1977. Ovalbumin gene is split in chicken DNA. *Nature* (London) 270:314–19.

Britten, R. J.; Povich, M.; and Smith, J. 1970. A new method for DNA purification. *Carnegie Institution Washington Yearbook*. Baltimore, Md.: Port City Press, 68:400–2.

Cairns, J. 1963. The chromosome of *Escherichia coli. Cold Spring Harbor Symp. Quant. Biol.* 28:43–46.

Chilton, M. D.; Drummond, M. H.; Merlo, D. J.; Sciaky, D.; Montoya, A. L.; Gordon, M. P.; and Nester, E. W. 1977. Stable incorporation of plasmid DNA into higher plant cells: the molecular basis of crown gall tumorigenesis. *Cell* 11:263–71.

Clarke, L., and Carbon, J. 1976. A colony bank containing synthetic ColEl Hybrid plasmids representative of the entire *E. coli* genome. *Cell* 9:91–99.

Cohn, R. H.; Lowry, J. C.; and Kedes, L. 1976. Histone genes of the sea urchin (*S. purpuratus*) cloned in *E. coli*: order, polarity, and strandedness of the five histone-coding and spacer regions. *Cell* 9:147–61.

Darby, G. K.; Jones, A. S.; Kennedy, J. F.; and Walker, R. T. 1970. Isolation and analysis of the nucleic acids and polysaccharides from *Clostridium welchii.J. Bacteriol.* 103:159–65.

DeWit, J. R.; Daniels, D.; Schroeder, J. L.; Williams, B. G.; Denniston-Thompson, K.; Moore, D. D.; and Blattner, F. R. 1980. Restriction maps for twenty-one charon vector phages. *J. Virol.* 33:401–10.

Early, P.; Huang, H.; Davis, M.; Calame, K.; and Hood, L. 1980. An immunoglobulin heavy chain variable region gene is generated from three segments of DNA: V_H, D and J_H. *Cell* 19:981–92.

Fritsch, E. F.; Lawn, R. M.; and Maniatis, T. 1980. Molecular cloning and characterization of the human β-like globin gene cluster. *Cell* 19:959–72.

Gerlach, W. L., and Bedbrook, J. R. 1979. Cloning and characterization of ribo-

somal RNA genes from wheat and barley. *Nucl. Acids Res.* 7:1869–85.

Grunstein, M., and Hogness, D. S. 1975. Colony hybridization: a method for the isolation of cloned DNAs that contain a specific gene. *Proc. Nat. Acad. Sci. (USA)* 72:3961–65.

Hardison, R. C.; Butler, E., III; Lacy, E.; Maniatis, T.; Rosenthal, N.; and Efstratiadis, A. 1979. The structure and transcription of four linked rabbit β-like globin genes. *Cell* 18:1285–97.

Heyn, R.; Hermans, A. K.; and Schilperoort, R. A. 1974. Rapid and efficient isolation of highly polymerized plant DNA. *Plant Sci. Lett.* 2:73–78.

Kaback, D. B., and Halvorson, H. O. 1977. Magnification of genes coding for ribosomal RNA in *Saccharomyces cerevisiae*. *Proc. Nat. Acad. Sci. (USA)* 74:1177–80.

Kacian, D. L., and Myers, J. C. 1976. Synthesis of extensive, possibly complete, DNA copies of poliovirus RNA in high yields and at high specific activities. *Proc. Nat. Acad. Sci. (USA)* 73:2191–95.

Katterman, F. R. H. 1975. Purification of DNA by cetyltrimethylammonium bromide as a potential source of error for the determination of 5-methylcytosine residues in DNA. *Anal. Biochem.* 63:156–60.

Lacy, E.; Hardison, R. C.; Quon, D.; and Maniatis, T. 1979. The linkage arrangement of four rabbit β-like globin genes. *Cell* 18:1273–83.

Lauer, J.; Shen, C. K. J.; and Maniatis, T. 1980. The chromosomal arrangement of human α-like globin genes: sequence homology and α-globin gene deletions. *Cell* 19:119–30.

Leder, P.; Tilghman, S. M.; Tiemeier, D. C.; Polisky, F. I.; Seidman, J. G.; Edgell, M. L.; Enquist, L. W.; Leder, A.; and Norman, B. 1977. The cloning of mouse globin and surrounding gene sequences in bacteriophage λ. *Cold Spring Harbor Symp. Quant. Biol.* 42:915–20.

Lifton, R. P.; Goldberg, M. L.; Karp, R. W.; and Hogness, D. W. 1977. The organization of the histone genes in *Drosophila melanogaster*: functional evolutionary implications. *Cold Spring Harbor Symp. Quant. Biol.* 42:1047–51.

Maniatis, T.; Hardison, R. C.; Lacy, E.; Lauer, J.; O'Connell, C.; Quon, D.; Sim, G. K.; and Efstratiadis, A. 1978. The isolation of structural genes from libraries of eucaryotic DNA. *Cell* 15:687–701.

Maniatis, T.; Jeffrey, A.; and Kleid, D. G. 1975. Nucleotide sequency of the rightward operator of phage λ. *Proc. Nat. Acad. Sci. (USA)* 72:1184–88.

Marmur, J. 1961. A procedure for the isolation of deoxyribonucleic acid from microorganisms. *J. Mol. Biol.* 3:208–18.

Max, E. E.; Seidman, J. G.; and Leder, P. 1979. Sequences of five potential recombination sites encoded close to an immunoglobulin x constant region gene. *Proc. Nat. Acad. Sci. (USA)* 76:3450–54.

Schaffner, W.; Kunz, G.; Daetwyler, H.; Tilford, J.; Smith, H. O.; and Birnstiel, M. L. 1978. Genes and spacers of cloned sea urchin histone DNA analyzed by sequencing. *Cell* 14:655–71.

Slightom, J.; Blechl, A. E.; and Smithies, O. 1980. Human fetal Gγ and Aγ globin genes: complete nucleotide sequences suggest that DNA can be exchanged between these duplicated genes. *Cell* 21:627–38.

Southern, E. M. 1975. Detection of specific sequences among DNA fragments separated by gel electrophoresis. *J. Mol. Biol.* 98:503–17.

Sternberg, N.; Tiemeier, D.; and Enquist, L. 1977. In vitro packaging of a λdam vector containing *Eco*RI DNA fragments of *Escherichia coli* and phage P1. *Gene* 1:255–80.

Sung, S. M.; Slightom, J. L.; and Hall, T. C. 1980. Three intervening sequences revealed by characterization and sequence comparison of cloned French bean phaseolin (G1-globulin) genomic and cDNAs. *Nature* (London) 289:37–41.

Thomashow, M. F.; Nutter, R.; Postle, K.; Chilton, M. -D.; Blattner, F. R.; Powell, A.; Gordon, M. P.; and Nester, E. W. 1980. Recombination between plant DNA and the Ti-plasmid of *Agrobacterium tumefaciens*. *Proc. Nat. Acad. Sci. (USA)* 77:6448–52.

Tonegawa, S.; Brack, C.; Hozumi, N.; and Schuller, R. 1977. Cloning of an immunoglobulin variable region gene from mouse embryo. *Proc. Nat. Acad. Sci. (USA)* 74:3518–22.

Vanin, E. F.; Goldberg, G. I.; Tucker, P.; and Smithies, O. 1980. A mouse alpha globin-related pseudo gene ($\psi\alpha$30.5) lacking intervening sequences. *Nature* (London) 286:222–26.

Walbot, V. and Goldberg, R. 1978. Plant genome organization and its relationship to classical plant genetics. In *Nucleic acids in plants*, ed. J. Davis and T. Hall, pp. 3–40. Boca Raton, Fla.: CRC Press.

Wang, J. C., and Kaiser, A. D. 1973. Evidence that the cohesive ends of mature λ DNA are generated by the gene A product. *Nature New Biol.* 241:16–17.

Watson, J. D. 1970. *Molecular biology of the gene*. New York: W. A. Benjamin.

4

THE CLONING AND ORGANIZATION
OF SOYBEAN LEGHEMOGLOBIN GENES

K. A. Marcker
K. Gausing
B. Jochimsen
P. Jørgensen
K. Paludan
E. Truelsen

Leghemoglobin is synthesized exclusively in the nitrogen-fixing root nodules that develop owing to the symbiotic association of bacteria (*Rhizobium*) with legumes. In soybean nodules there are three major forms of leghemoglobin, termed leghemoglobin a, c_1, and c_2, respectively. The difference between leghemoglobin a and c is six amino acids, while leghemoglobin c_1 and c_2 differ in only one amino acid (Sievers, Huhtala, and Ellfolk 1978). The ability of leghemoglobin to bind oxygen creates a very low oxygen tension within the nodules, thus protecting the oxygen-sensitive nitrogenase against denaturation. Despite the low oxygen tension oxidative phosphorylation is still able to proceed because leghemoglobin binds oxygen reversibly. Consequently, for effective nitrogen fixation to occur leghemoglobin must be present within the nodule. Leghemoglobin has never been detected in plants not fixing nitrogen. It is quite possible that the gene for leghemoglobin is not present in these plants. Should this be the case, it follows that attempts to transfer nitrogenase genes to such plants in order to generate new nitrogen-fixing combinations invariably also must involve the transfer of leghemoglobin genes.

In soybean nodules leghemoglobin synthesis may account for about 20 percent of the total protein synthesis. Leghemoglobin is encoded in a poly(A)$^+$ 9s messenger ribonucleic acid (mRNA) species, which is translated on cytoplasmic ribosomes (Sidloi-Lumbroso and Schulman 1977). Recent evidence (Baulcombe and Verma 1978, Sidloi-Lumbroso, Kleinmann, and

Schulman 1978) strongly suggests that leghemoglobin is a plant gene product. Since the gene is never active without prior infection with *Rhizobium*, it follows that in this particular case a higher eukaryotic gene somehow must be under the control of a procaryotic function.

This chapter reports the cloning and characterization of a synthetic deoxyribonucleic acid (DNA) copy of soybean leghemoglobin mRNA. In addition, we describe some preliminary analyses of the corresponding chromosomal genes.

SYNTHESIS AND CLONING OF STRUCTURAL GENE SEQUENCES

A double-stranded DNA copy of soybean leghemoglobin mRNA was synthesized and cloned in the bacterial plasmid pBR322 in the following way (for details, see Truelsen et al. 1979). Polysomes were prepared from root nodules of soybean (var. Bråvalla) 4 weeks after infection with *Rhizobium japonicum* strain 61A69. The polysomes were digested with proteinase K, and poly(A)$^+$ mRNA was isolated by chromatography on an oligo-dT cellulose column. The poly(A) containing mRNA was fractionated on a 5 to 20 percent sucrose gradient. The 9-10s fraction corresponding to leghemoglobin mRNA was used for synthesis of complementary DNA (cDNA) using the enzyme avian myoblastosis virus reverse transcriptase.

Double-stranded DNA (dsDNA) was synthesized by DNA polymerase with cDNA as template, and the resulting product was treated with S1 nuclease. After isolation of the dsDNA, poly(dA) tails were added using terminal transferase. Poly(dT) tails were added to the ends of plasmid pBR322 linearized by digestion with the restriction endonuclease *Bam*HI. The tailed plasmid and the tailed dsDNA were annealed. After annealing the incubation mixture was used for transformation of *Escherichia coli* K12 strain HB101. Plasmid pBR322 contains both an ampicillin resistance gene and a tetracycline resistance gene.

The integration of foreign DNA sequences into the *Bam*HI site of pBR322 leads to inactivation of the tetracycline resistance gene, while the ampicillin resistance gene is not affected. Ampicillin-resistant but tetracycline-sensitive colonies were screened by colony hybridization using purified [125I]-labeled dsDNA as a hybridization probe. Plasmid DNA was isolated from positive colonies. The sizes of the insertions were estimated by gel electrophoresis in agarose gels using pBR322 and pBR322 with a 700 base-pair insertion as markers. Two clones, 24-31 and 7-56, carrying insertions of about 700 base pairs were selected for further analysis.

ANALYSIS OF CLONED SEQUENCES

Plasmid DNA from these two clones were digested with a variety of restriction endonucleases. The complete DNA sequence of pBR322 has recently been published (Sutcliffe 1978). This allows rather accurate sizing of the various restriction fragments obtained. On the basis of these data, a restriction cleavage map for the inserted leghemoglobin DNA could be constructed as shown in Fig. 4.1.

In the inserted DNA there is no cleavage site for *EcoRI*, *SalI*, *BamHI*, *AvaI*, *KpnI*, *HaeII*, *BglII*, *SmaI*, *XhoI*, *XbaI*, *SacI*, *HpaI*, *HpaII*, and *ClaI*. The inserted DNA contains single cleavage sites for *HindIII*, *PstI*, *HindII*, *HinfI*, *HaeIII*, *Sau3A*, and *TaqI* and several sites for *AluI*. The amino acid sequence of the major forms of leghemoglobin has recently been elucidated (Sievers, Huhtala, and Ellfolk 1978). The relative positions of the restriction sites are in accordance with the amino acid sequence. The relation between the location of the restriction sites and the corresponding amino acid is shown in Table 4.1.

Finally, the orientation of the inserted DNA could be deduced with the result that the 3' end of leghemoglobin RNA is closer to the *EcoRI* site of pBR322 than the 5' terminal end.

Final confirmation that the cloned DNA corresponds to leghemoglobin DNA has been obtained by DNA sequence analysis, which was performed according to the methods of Maat and Smith (1978) and Maxam and Gilbert (1977). Almost the entire base sequence of the inserted DNA has been determined. There is complete agreement between the amino acid sequence and the DNA sequence except in two cases. Sievers, Huhtala, and Ellfolk (1978) determined the sequence -Phe-Val-Val-Lys- in positions 102 to 105, while we find a sequence corresponding to -Phe-Val-Val-Val-Lys- in clone 7-56. The -COOH terminal found by DNA analysis is *Ala*. Sievers, Huhtala, and Ellfolk (1978) found *Phe* or *Lys* as -COOH terminals. They also detected *Ala* as a -COOH terminal but failed to find the corresponding -COOH terminal peptide and consequently concluded that the determination of *Ala* as a -COOH terminal was due to an artifact. Our data clearly show that this is not so and that a leghemoglobin species having an *Ala* as -COOH terminal must exist. The DNA sequence further shows that the inserted DNA in 7-56 is a copy of leghemoglobin c mRNA.

Clone 24-31 does not have the *Alu* site corresponding to amino acids -Glu-Ala- at positions 114-115. This corresponds to a difference between leghemoglobin c and a, and, therefore, the inserted sequence in clone 24-31 corresponds most likely to leghemoglobin a. The 3' non-coding end is about 135 nucleotides long. The stop codon is UAA, and there is an additional UAA codon in phase 15 bases further downstream. The insertion in clones

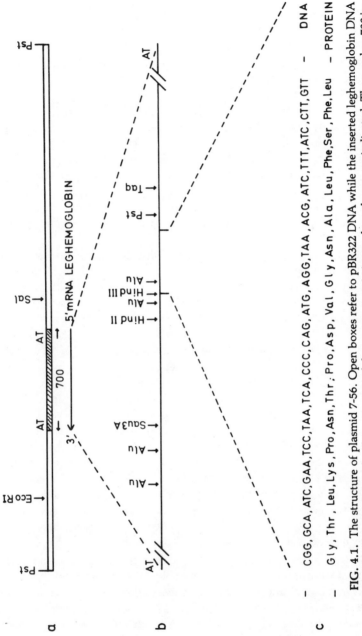

FIG. 4.1. The structure of plasmid 7-56. Open boxes refer to pBR322 DNA while the inserted leghemoglobin DNA is indicated by the dark box. The cleavage sites of the various restriction endonucleases are indicated. The number 700 indicates that the inserted DNA is about 700 base pairs long including the poly(dAT) tails. (a) Linear cleavage map of plasmid 7-56. The orientation of the inserted leghemoglobin DNA is indicated below the box. (b) Detailed structure of the inserted leghemoglobin DNA. (c) The base sequence of part of the inserted DNA compared with the amino acid sequence from residue 42 to 60 in leghemoglobin from soybeans.

c

CGG, GCA, ATC, GAA, TCC, TAA, TCA, CCC, CAG, ATG, AGG, TAA , ACG, ATC, TTT, ATC, CTT, GTT   —   DNA

Gly, Thr, Leu, Lys, Pro, Asn, Thr ; Pro, Asp, Val, Gly, Asn, Ala , Leu, Phe, Ser, Phe, Leu   —   PROTEIN

TABLE 4.1
Correspondence Between Restriction Endonuclease Cleavage Sites in the Insertion of Clone 7-56 and Amino Acid Sequence of Leghemoglobin.

Restriction Enzyme	Amino Acid Sequence
PstI	-Pro-Ala-Ala- (38,39,40)
HindIII	-Lys-Leu- (64,65)
HindII	-Gly-Glu-Leu- (74,75,76)
TaqI	-Ser-Ile- (32,33)
HaeIII	-Gly-His- (60,61)
HinfI	-Asp-Ser- (71,72)
Sau3A	-Asp-Pro- (99,100)
AluI	-Lys-Leu- (57,58)
	-Lys-Leu- (64,65)
	-Ser-Ala- (72,73)
	-Glu-Ala- (114,115)
	-Ala-Ala- (137,138)

Note: All cleavage sites have been confirmed by DNA sequence analyses.
Source: Compiled by the authors.

7-56 and 24-31 does not include the entire structural sequence but lacks the sequences corresponding to the first 10 to 15 amino acids. However, we have recently cloned the leghemoglobin RNA sequences again using the polyGC tailing method. Analyses of these clones show that in several cases the entire structural sequence has been cloned.

ANALYSIS OF CHROMOSOMAL LEGHEMOGLOBIN GENES

Chromosomal soybean DNA was prepared from 4-day-old seedlings that were germinated in the dark to minimize contamination with chloroplast DNA. The DNA was finally purified in a CsCl gradient containing 10 µg/ml of ethidium bromide. In analogy with others we have found that plant DNA is partially resistant to digestion with several restriction endonucleases, presumably due to a rather extensive methylation. However, EcoRI and XbaI digested soybean DNA. Soybean DNA digested with either of these enzymes was subjected to gel electrophoresis and transferred onto nitrocellulose filters according to the procedure of Southern (1975). Hybridization was carried out using nick-translated plasmid DNA from clone 7-56. Nick translation of DNA was according to Rigby et al. (1977) and usually resulted in a specific activity of 4-5 × 10⁷ cpm/µg DNA.

Prehybridization and hybridization of filters were as described by Jeffreys and Flavell (1977) except that Dextran sulphate was added to a final concentration of 10 percent w/v. After extensive washing the filters were autoradiographed, usually for about 4 to 6 days.

The amino acid sequence data for leghemoglobin indicate the existence of at least three separate genes for leghemoglobin in soybeans. Hybridization data suggest the presence of up to 40 gene copies per genome (Baulcombe and Verma 1978).

Our hybridization data for soybean DNA restriction nuclease fragments with nick-translated 7-56 DNA are given in Table 4.2, and an example of such experiments is shown in Fig. 4.2. The data are not sufficient to determine the exact number of genes per genome. However, the *Eco*RI data show the existence of at least six fragments hybridizing with about equal intensity to complementary sequences. Among several possible models we would like to discuss two models in greater detail in order to explain our data.

Model A. This model predicts that there are six (or at least six) separate leghemoglobin genes.

Model B. This model predicts only three separate genes. *Eco*RI does not cleave the coding sequence, and the six *Eco*RI fragments must therefore indicate an *Eco*RI site within an intron sequence in all three genes.

Digestion with *Xba*I results in the detection of three fragments with molecular weights corresponding to 29, 13, and 12 kb and hybridization

TABLE 4.2
Sizes of Soybean DNA Restriction Fragments Hybridizing to Clone 7-56 (in kb).

*Xba*I	*Eco*RI	*Xba*I/*Eco*RI
29.0	12.5	11.5
13.0	11.5	8.5
11.2	7.5	7.5
	6.0	6.0
	5.5	5.5
	4.3	4.3

Note: The accuracy of the size determination is about ±5 percent, except for the 29 kb fragment where the accuracy is on the order of about ±10 percent.
Source: Compiled by the authors.

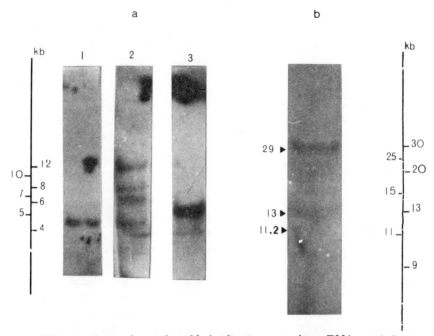

a   b

FIG. 4.2. Autoradiography of hybridization to soybean DNA restriction endonuclease fragments with radioactive leghemoglobin DNA sequences. (a) Soybean DNA (lane 2) was digested with EcoRI and electrophoresed in a 1.25 percent agarose gel (120 v, 4 hours). Hybridization controls consisted of linearized pBR322 (lane 1) and linearized DNA from clone 7-56 (lane 3). An additional size marker was EcoRI digested DNA from bacteriophage λ. DNA was transferred to nitrocellulose filters and hybridized to nick-translated 7-56 with a specific activity of 1.3 × 10^8 cpm/μg. Autoradiography was performed with an intensifying screen at −70°C for 4 days. (b) Soybean DNA was digested with XbaI (BioLabs) and electrophoresed in a 0.7 percent agarose gel. Hybridization and autoradiography as in a.

densities of approximately 4:1:1, respectively. This is compatible with model A. The 29 kb XbaI fragment most likely contains four genes. The two unallocated genes are then situated on the two smaller Xba fragments. An alternative in model A is that the 29 kb fragment in fact consists of several fragments comigrating at the 29 kb position, each representing one or more leghemoglobin genes. However, we have never been able to resolve this fragment into more than one band — despite running it in several different agarose gel systems.

In model B it is predicted that there exist three separate genes for leghemoglobin in soybean and that each of these genes has an EcoRI cleavage site in an intron. The difference in the hybridization intensities of the *Xba* fragments are not easily explained by this model. However, we feel that this possibility cannot entirely be ruled out. Based on the available data, model A is the most plausible, although we are aware that alternative models are possible.

In order to investigate the chromosomal leghemoglobin genes in more detail, we have constructed a library of soybean genes using a lambda vector. So far 150×10^3 clones have been screened for leghemoglobin sequences. Positive clones are now under investigation.

ADDENDUM

Five different chromosomal leghemoglobin genes have now been cloned. One gene that is contained in a 7.5 kb soybean DNA fragment has been investigated in detail. DNA sequence analysis has shown that the isolated gene is a leghemoglobin *c* gene. Three intervening sequences interrupt the coding sequence of the isolated leghemoglobin gene at codons 32, 68-69, and 103-104. The sizes of the intervening sequences are 165 bp, ~225 bp, and ~300 bp, respectively. The locations of the splicing points of the first and third intervening sequences in the leghemoglobin gene are identical to the positions of the two interruptions found in all investigated globin genes, while the central intervening sequence is unique for leghemoglobin. The DNA sequences at the 5' and 3' ends of the intervening sequences are in good agreement with the consensus sequence for splicing junctions. Thus plant genes share this feature with other eukaryotic genes.

In conclusion our results support the notion that leghemoglobin and all other globins have a common evolutionary origin because of the identical positions of two intervening sequences. Leghemoglobin and myoglobin are generally assumed to have evolved earlier from the common ancestral gene than α and β globins. Thus it is a reasonable assumption that the central intervening sequence in the leghemoglobin gene was also present in the ancestral globin gene and subsequently was lost during its evolution into α and β globins.

REFERENCES

Baulcombe, D., and Verma, D. P. S. 1978. Preparation of a complementary DNA for leghemoglobin and direct demonstration that leghemoglobin is encoded by the soybean genome. *Nucl. Acids Res.* 5:4141–53.

Jeffreys, A. J., and Flavell, R. A. 1977. A physical map of the DNA region flanking the rabbit β-globin gene. *Cell* 12:429–39.

Maat, J., and Smith, A. J. H. 1978. A method for sequencing restriction fragments with deoxynucleotide triphosphates. *Nucl. Acids Res.* 5:4537–45.

Maxam, A. M., and Gilbert, W. 1977. A new method for sequencing DNA. *Proc. Nat. Acad. Sci. (USA)* 74:560–64.

Rigby, W. J. P.; Dieckmann, M.; Rhodes, C.; and Berg, P. 1977. Labelling deoxyribonucleic acid to high specific activity *in vitro* by nick translation with DNA polymerase I. *J. Mol. Biol.* 113:237–51.

Sidloi-Lumbroso, R.; Kleinmann, L.; and Schulman, H. M. 1978. Biochemical evidence that leghemoglobin genes are present in the soybean but not in *Rhizobium* genome. *Nature* 273:558–60.

Sidloi-Lumbroso, R., and Schulman, H. M. 1977. Purification and properties of soybean leghemoglobin messenger RNA. *Biochim. Biophys. Acta* 476:295–302.

Sievers, G.; Huhtala, M.-L.; and Ellfolk, N. 1978. The primary structure of soybean (*Glycine max*) leghemoglobin *c*. *Acta Chem. Scand. B* 32:380–86.

Southern, E. M. 1975. Detection of specific sequences among DNA fragments separated by gel electrophoresis. *J. Mol. Biol.* 98:503–17.

Sutcliffe, J. G. 1978. pBR322 restriction map derived from the DNA sequence: accurate DNA size markers up to 4361 nucleotide pairs long. *Nucl. Acids Res.* 5:2721–28.

Truelsen, E.; Gausing, K.; Jochimsen, B.; Jorgensen, P.; and Marcker, K. A. 1979. Cloning of soybean leghemoglobin structural gene sequences synthesized *in vitro*. *Nucl. Acids Res.* 6:3061–72.

5

CLONING OF DNA SEQUENCES
CODING FOR ZEIN PROTEINS
OF MAIZE

G. Feix
P. Langridge
U. Wienand

INTRODUCTION

The successful application of molecular cloning has led recently to new insights into genome structure and its expression in animal systems (Axel, Maniatis, and Fox 1979). However, comparable studies with plant systems are only beginning (Leaver 1980). In addition to yielding new information on plant genomes, such studies may be a necessary prerequisite for future attempts at improving plant properties by genetic engineering.

Deoxyribonucleic acid (DNA) sequences coding for proteins have been cloned either directly from chromosomal DNA (genomic clones) or after previous construction of double-stranded DNA structures from the messenger ribonucleic acids (mRNAs) of interest (complementary DNA [cDNA] clones) (Wu 1979). The preparation of cDNA clones of a particular system is usually attempted initially since these clones can then be used to identify genomic clones. cDNA clones have also proved to be helpful tools in the elucidation of some aspects of the mechanism of gene expression.

The construction of cDNA clones is greatly facilitated if the mRNAs occur in great amounts and can be isolated in an enriched form. This has been demonstrated for the production of globin and other cDNA clones in animals and fungi (Efstratiadis and Villa-Komaroff 1979). In plants high concentrations of specific mRNAs occur during storage protein synthesis in developing seeds. Storage proteins are often synthesized in large amounts, with high tissue specificity and at defined times of development, accompanied by the accumulation of the corresponding mRNAs. This advantageous situation has already been exploited for the preparation of cDNA clones for

73

globulin of beans (Hall et al. 1980) and for the hordeins of barley (Brandt 1979). The first storage protein system, however, to be used for the construction of cDNA clones was that of maize endosperm where the zein proteins are synthesized (Wienand, Brüschke, 19 and Feix 1979).

Zein proteins consist of two major components (19,000 M_r and 22,000 M_r) coded by separate mRNAs, which represent a major proportion of the developing endosperm mRNAs. Therefore, a total endosperm mRNA preparation has been employed as the starting material for the construction of cDNA clones specific for zein coding sequences. This procedure also led to the construction of a collection of non-zein endosperm cDNA clones. These endosperm clones are promising, since many functions of the developing endosperm of maize have been studied biochemically and genetically and are awaiting further molecular analysis (Nelson 1978).

In the following pages some aspects of the preparation of the cDNA clones as well as possible applications of the zein-specific clones are discussed.

PREPARATION OF cDNA CLONES

Initially, cDNA was synthesized with reverse transcriptase from an RNA preparation isolated from 21 days postpollination endosperm. This RNA represents a population of poly(A)-mRNAs active in translation at this time in endosperm development (Wienand and Feix 1978). The electrophoretic separation of this RNA under denaturing conditions allows an estimation of the size and homogeneity of the zein mRNAs (see Fig. 5.1, lane **A**). The predominant RNA bands seen in this gel are the ribosomal RNAs (rRNAs). Although in vitro translation of such an RNA preparation leads to an extensive synthesis of zein proteins, no distinct band of zein-specific RNA can be seen in the electrophoretic separation profile. The region of the gel containing zein mRNAs (indicated in the figure) was identified by the in vitro translation of RNA fractions eluted from individual gel slices. The approximate size of the zein mRNAs can be estimated by comparison with the running positions of the marker RNAs shown in the other lanes of Fig. 5.1. The zein mRNA-containing fractions were used for the selection of zein-specific clones (see below). This approach of using specific RNA fractions for the identification of clones may be applicable for the selection of further cDNA clones from the available endosperm-specific clones.

It should be noted that the use of poly(A) mRNA isolated from protein bodies gives an RNA fraction particularly rich in zein mRNAs and, hence, is useful for the preparation of zein-specific cDNA clones (Burr 1979). Protein bodies contain at their surface polysomes actively synthesizing zein (Burr and Burr 1976).

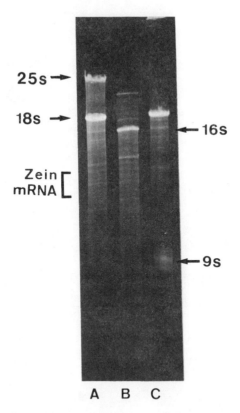

FIG. 5.1. Urea polyacrylamide gel electrophoresis of RNA. Poly(A) poly-somal RNA from maize endosperm 21 days postpollination (lane **A**) and various marker RNAs (lanes **B** and **C**) were electrophoresed under denaturing conditions according to the procedure of Wienand and Feix (1978). Lane **B** = ribosomal RNA from *E. coli*; Lane **C** = polysomal RNA from rabbit reticulocytes.

The method used to convert cDNA to double-stranded DNA and to subsequently anneal it into plasmid pBR322 is represented in Fig. 5.2. Details of this procedure can be found in the recent publication of Wienand, Brüschke, and Feix (1979). Clones containing recombinant plasmids, as recognized by their resistance to tetracycline and their sensitivity to ampicillin, were screened by colony hybridization assays with [^{125}I]-labeled RNA and cDNA probes. Poly(A) polysomal RNA isolated from endosperm or cDNA derived from it hybridized to approximately 25 percent of the total clones. Hybridization with zein-enriched mRNA (see above) led to the identification of zein-specific clones amounting to 8 percent of the total number of clones. Fig. 5.3 shows an example of the colony hybridization test with zein mRNA as hybridization probe.

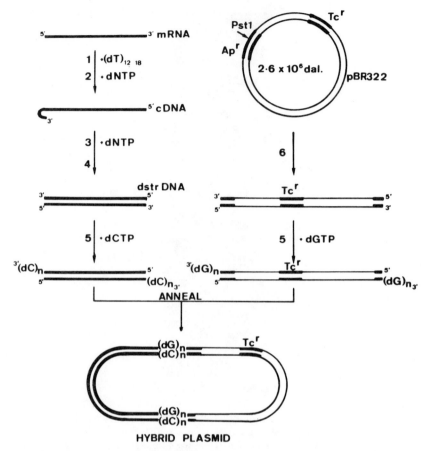

FIG. 5.2. Schematic representation of the in vitro synthesis of hybrid plasmids. The numbers stand for the following reagents: **1** = reverse transcriptase; **2** = 0.25 N NaOH; **3** = DNA polymerase I; **4** = S1 nuclease; **5** = terminal transferase; and **6** = restriction endonuclease *Pst*I.

It is interesting to note that a colony hybridization assay with [125I]-labeled maize rRNA did not indicate the presence of clones containing rRNA sequences. This confirms that maize rRNA will not serve as template for cDNA synthesis under the assay conditions employed here. Conversely, soybean rRNA was recently shown to direct the synthesis of cDNA (Sullivan, Brisson, and Verma 1980).

The zein-specific clones were clearly identified by in vitro translation of RNAs that hybridized specifically to cloned DNAs preselected as zein specific by the colony hybridization experiments. Zein proteins synthesized

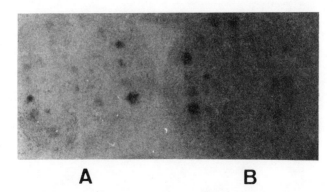

A   B

FIG. 5.3. Autoradiography of colony hybridization test. Poly(A) polysomal RNA, enriched for zein mRNA by separation on a polyacrylamide gel, was used as a hybridization probe to locate zein cDNA clones by the colony hybridization method. Two representative filters are shown.

by these RNAs were identified by precipitation with zein-specific antibodies and were subsequently analyzed by SDS-polyacrylamide electrophoresis (Wienand and Feix 1978). Antibody identification was particularly important, as zein proteins do not exhibit any known enzymatic activity. Fig. 5.4 shows a representative example of such an analysis for three different cDNA clones. Approximately half the zein mRNA sequences were represented in these clones. Under the conditions used, the clones hybridize specifically either with mRNA coding for the 19,000 M_r protein (lanes **D** and **E**) or with mRNA coding for the 22,000 M_r protein (lanes **F** and **G**). In addition to these two classes of zein-specific clones, a third class hybridized to mRNAs coding for the 19,000 M_r protein and for another zein protein of 20,000 M_r (lane **H**). The 20,000 M_r protein has consistently been found as a product of the in vitro translation of endosperm RNA and of protein body RNA, although in relatively low amounts.

Work is in progress to analyze with the help of the cDNA clones whether this zein protein is coded for by a separate mRNA. It is noteworthy that the recent analysis of another set of zein-specific cDNA clones — obtained by starting from protein body RNA and by using the A/T connector method — did not show any clone of the third class (Park, Lewis, and Rubenstein 1980). The reasons for this discrepancy may reside in the different conditions used for hybridization or in vitro translation or may result from differences of the zein-specific sequences contained in the two sets of clones. It appears significant that 20 percent of the zein-specific clones corresponded to the third class, while 40 percent belonged to each of the other two classes.

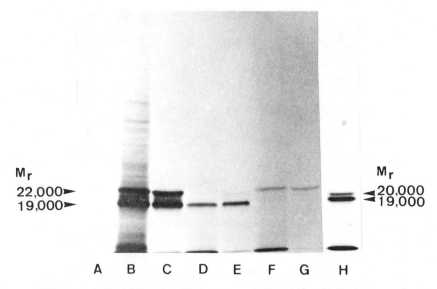

FIG. 5.4. SDS-polyacrylamide gel electrophoresis of in vitro synthesized proteins. RNA prepared from developing maize seed was hybridized to the cDNA clones and the hybridized mRNAs eluted and translated in the wheat germ in vitro system (Wienand and Feix 1978). The products of translation were separated by SDS-polyacrylamide gel electrophoresis and revealed by fluorography. The fluorography shows the in vitro translation product: without added mRNA (lane A), with polysomal mRNA from endosperm (lanes B and C), with mRNA hybridized to plasmid of class I (D and E), with mRNA hybridized to plasmid of class II (F and G), and with mRNA hybridized to plasmid of class III (H). The products in lanes C, E, G, and H were immunoprecipitated with rabbit antizein antibodies prior to electrophoresis.

HYBRIDIZATION ANALYSIS

A major asset of cDNA clones is their great potential as specific hybridization probes for the analysis fo cellular nucleic acid structures. For example, little information is available about the concentration of zein mRNA in developing endosperm and the nature of the primary transcription products of the zein genes, particularly for the opaque and similar mutants, with reduced levels of some zein proteins. In this respect, the "northern" technique involving RNA/DNA hybridizations will be particularly useful (Alwine, Kemp, and Stark 1977). Fig. 5.5 demonstrates the use of this technique for detecting one or more of the major zein mRNA classes. The zein-specific sequences of the plasmids hybridize only with their homologous mRNAs,

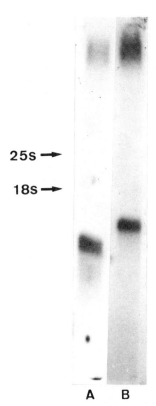

25s ➞

18s ➞

FIG. 5.5. Formaldehyde-agarose gel electrophoresis of RNA. Poly(A) polysomal RNA from
endosperm 21 days postpollination was used for
"northern" type analysis based upon the procedure of
Alwine, Kemp, and Stark (1977). [^{32}P]-labeled
hybrid plasmid specific for the 19,000 M_r protein was
hybridized to poly(A)-RNA (lane **A**). This label was
then removed by washing, and [^{32}P]-labeled plasmid
specific for the 22,000 M_r protein was hybridized in
the same filter (lane **B**). The positions of the major
rRNAs are indicated.

A   B

which had been previously separated by gel electrophoresis under denatur-
ating conditions.

The zein-specific inserts of approximately 50 percent of the recombinant plasmids can be excised by digestion with the restriction enzyme *Pst*I.
This has proved helpful when using the cloned sequences as probes in
Southern-type experiments (Southern 1975) to analyze zein coding regions
of the maize genome. Restriction enzyme fragments of total maize DNA,
prepared from seedlings, were electrophoresed in 0.8 percent agarose, transferred to nitrocellulose paper, and hybridized with [^{32}P]-labeled cloned insert fragments. It was essential to use inserts rather than the total plasmid
DNA to obtain consistent results in the Southern experiments.

The autoradiography of a typical banding pattern is given in Fig.5.6.
Maize DNA contains at least 12 *Bam*HI fragments in the size range from
$1.5 \times 10^6 M_r$ to $12 \times 10^6 M_r$ (lanes **A** and **B**) and 9 *Eco*RI fragments in the
size range from $2.5 \times 10^6 M_r$ to $12 \times 10^6 M_r$ (lanes **C** and **D**) that hybridize with cloned insert DNA specific for the 19,000 M_r zein proteins.

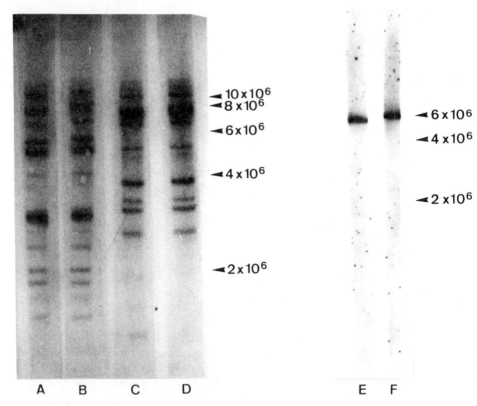

FIG. 5.6. Agarose gel electrophoresis of zein-specific restriction enzyme fragments of maize DNA. Maize DNA was digested with *Eco*RI or *Bam*HI; the fragments were separated by gel electrophoresis and transferred to nitrocellulose membrane. Lanes A, B, C, and D represent hybridizations with insert fragments containing sequences coding for the 19,000 M_r zein protein. Lanes E and F represent a hybridization with maize rRNA. Lanes C, D, E, and F contain *Eco*RI fragments; lanes A and B, *Bam*HI fragments. Lanes A and B, and C and D as well as E and F represent duplicate experiments. Zein-specific sequences were revealed on the membranes by hybridization with radioactively labeled cloned zein cDNA and autoradiography.

Lanes E and F of Fig. 5.6 represent the result of hybridizing [^{125}I]-labeled rRNA to *Eco*RI fragments. The single hybridization band of an approximate M_r of 6×10^6 obtained with rRNA is a strong indication for the complete digestion of the maize DNA with *Eco*RI. Further bands were revealed by hybridization with the other classes of zein cDNA clones (for details see Wienand and Feix [1980]). These results indicate the presence of a multigene system coding for the zein proteins, since one can see not only many bands in the Southern analysis but also variations in the intensity of labeling (weaker bands represent single genome copies). It is possible that this Southern labeling pattern seen using seedling DNA may vary with DNA from different tissues, particularly endosperm.

The multiplicity of zein genes as indicated by the Southern analysis has become increasingly important with the variation in recently published estimates of the number of zein genes (between 5 and 120) per haploid genome (Pedersen et al. 1980; Viotti et al. 1979). A further important feature of this issue relates to the improvement in protein quality of endosperm protein by genetic engineering of structural genes. For example, the endosperm protein is nutritionally deficient in lysine (Gianazza et al. 1977).

CONCLUSION

Correlation of the multiple hybridization bands to a defined number of zein genes will only be possible with a knowledge of the restriction enzyme map of zein genome structures and after an analysis of the genes for a possible split-gene structure. These studies are most efficiently undertaken with genomic zein clones. However, the cDNA clones are of value not only in providing a ready means for identifying genomic clones but also in investigating the expression of the zein genes. Studies on mRNA levels during development and in various maize mutants will be dependent upon the cDNA clones. It is also hoped to use the clones to examine the possibility of processing the initial zein mRNA transcripts and the coordination of synthesis of the different zein proteins.

The advantages of the zein system as a model system for the investigation of basic questions of plant genomic structure have been further enhanced by the availability of zein-specific clones.

ADDENDUM

The cDNA clones have been used to select a genomic zein clone from a collection of recombinant λ phages. The recombinant phages were prepared in the λgtWES cloning system from EcoRI restriction enzyme fragments of maize DNA. The isolated clone contains a 4.4 kb maize DNA fragment, and the zein gene (0.8 kb) is located near the center of the fragment. The electron micrograph (Fig. 5.7) shows an R-loop structure formed between the 4.4 kb fragment and the zein mRNA. No intervening sequences were detected by the R-loop analysis. The cloned DNA fragment was found, by Southern blot analysis, to correspond to a maize EcoRI DNA fragment coding for the 19,000 dalton zein protein. The central location of the zein gene, the

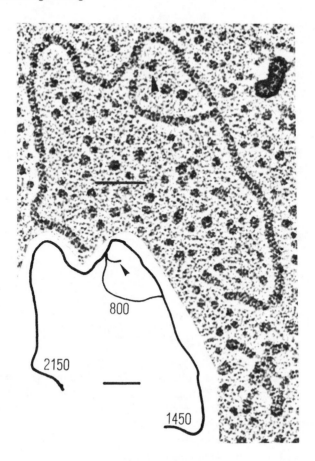

FIG. 5.7. Electron micrograph and line drawing of an R-loop structure between zein mRNA and the cloned maize DNA fragment. The bar represents 0.1 μm, and the size of the DNA sections is indicated in base pairs. The poly(A) tail of the zein mRNA is marked by arrow.

absence of major intervening sequences, and the size of the cloned maize DNA fragment make this clone attractive for further studies. (Wienand, Langridge, and Feix 1981).

The cDNA clones have been used as hybridization probes in northern blot experiments to identify several zein mRNA precursor species, of 2.4 kb and greater, in RNA preparations from 19- to 22-day-old endosperm. Northern experiments also showed that differences in zein protein accumulation during the development of wild type and opaque kernels coincided with the amount and type of zein mRNAs present in the endosperm. This result suggests that maize endosperm does not contain a specialized protein synthetic apparatus that is modified in the opaque mutation (Burr and Burr 1979).

REFERENCES

Alwine, J. C.; Kemp, D. J.; and Stark, G. R. 1977. Method for detection of specific RNAs in agarose gels by transfer to diazobenzyloxymethyl-paper and hybridization with DNA probes. *Proc. Nat. Acad. Sci. (USA)* 74:5350–54.

Axel, R.; Maniatis, T.; and Fox, F. C., eds. 1979. Eucaryotic gene regulation. *ICN-UCLA Symposia on Molecular and Cellular Biology*, vol. 14. New York: Academic Press.

Brandt, A. 1979. Cloning of double-stranded DNA coding for hordein polypeptides. *Carlsberg Res. Commun.* 44:255–67.

Burr, B., and Burr, F. A. 1976. Zein systhesis in maize endosperm by polyribosomes attached to protein bodies. *Proc. Nat. Acad. Sci. (USA)* 73:515–19.

Burr, B. 1979. Identification of zein structural genes in the maize genome. In *Seed protein improvement in cereals and grain legumes*, vol. 1. Vienna: International Atomic Energy Agency.

Burr, F. A., and Burr, B. 1979. Molecular basis of zein protein synthesis in maize endosperm. In *The Plant Seed*, ed. I. Rubinstein, R. L. Phillips, C. E. Green, and B. G. Gengenbach, pp. 27–48. New York: Academic Press.

Efstratiadis, A., and Villa-Komaroff, L. 1979. Cloning of double-stranded cDNA. In *Genetic engineering, principles and methods*, ed. J. C. Setlow and A. Hollaender, 1:15–36. New York: Plenum Press.

Gianazza, E.; Viglienchi, V.; Righetti, P. G.; Salamini, F.; and Soave, C. 1977. Amino acid composition of zein molecular components. *Phytochemistry* 16:315–17.

Hall, T. C., Sun, S. M.; Buchbinder, B. U.; Pyne, J. W.; Bliss, F. A.; and Kemp, J. D. 1980. Bean seed globulin mRNA: translation, characterization and its use as a probe towards genetic engineering of crop plants. In *Genome organization and expression in plants*, ed. J. C. Leaver, pp. 259–72. New York: Plenum Press.

Leaver, C. J., ed. 1980. Genome organization and expression in plants. *NATO Advanced Study Institute Series: Series A., Life Sciences*, vol. 29. New York: Plenum Press.

Nelson, O. E. 1978. Gene action and endosperm development in maize. In *Maize breeding and genetics*, ed. B. D. Walden, pp. 389–403. New York: John Wiley.

Park, W. D.; Lewis, E. D.; and Rubenstein, I. 1980. Heterogeneity of zein mRNA and protein in maize. *Plant Physiol.* 65:98–106.

Pedersen, K.; Bloom, K. S.; Anderson, J. N.; Glover, D. V.; and Larkins, B. A. 1980. Analysis of the complexity and frequency of zein genes in the maize genome. *Biochemistry* 19:1644–50.

Southern, E. M. 1975. Detection of specific sequences among DNA fragments separated by gel electrophoresis. *J. Mol. Biol.* 98:503–17.

Sullivan, D. E.; Brisson, N.; and Verma, D. P. S. 1980. Reverse transcription of

25s soybean ribosomal RNA in the absence of exogenous primer. *Biochem. Biophys. Res. Commun.* 94:144–50.

Viotti, A.; Sala, E.; Marotta, R.; Alberi, P.; Balducci, C.; and Soave, C. 1979. Genes and mRNAs coding for zein polypeptides in *Zea mays. Eur. J. Biochem.* 102:211–22.

Wienand, U.; Brüschke, C., and Feix, G. 1979. Cloning of double-stranded DNAs derived from polysomal mRNA of maize endosperm: isolation and characterization of zein clones. *Nucl. Acids Res.* 6:2707–15.

Wienand, U., and Feix, G. 1978. Electrophoretic fractionation and translation *in vitro* of poly(rA)-containing RNA from maize endosperm. *Eur. J. Biochem.* 92:605–11.

——— . 1980. Zein specific restriction enzyme fragments of maize DNA. *FEBS Lett.* 116:14–16.

Wienand, U., Langridge, P., and Feix, G. 1981. Isolation and characterization of a genomic sequence coding for a zein gene. *Molec. gen. Genet.*, in press.

Wu, R., ed. 1979. *Recombinant DNA.* Methods in enzymology, ed. S. P. Collowick and N. O. Kaplan, vol. 68. New York: Academic Press.

6

THE SINGLE-STRANDED DNA PLANT VIRUSES

Robert M. Goodman

Among the plant viruses the majority contain genomes of single-stranded ribonucleic acid (RNA). Most of these contain the messenger-sense RNA. The phytoreoviruses contain segmented genomes of double-stranded RNA. The caulimoviruses (see Chapter 7) are the only plant viruses with double-stranded deoxyribonucleic acid (dsDNA) and for 10 years were the only plant viruses known that contained DNA (Shepherd 1979).

A new group of DNA plant viruses was discovered in 1977. Now called geminiviruses (Matthews 1979), they are the first and only group of plant viruses known containing single-stranded DNA (ssDNA). The geminiviruses occur in nature primarily in the tropics, although one probable member of the group is found in temperate areas of the western United States (Goodman 1980). The diseases they cause are among the most important diseases of a number of major tropical food and fiber crops. Recent reviews have concerned the epidemiology, agricultural importance, and diagnostic aspects of geminiviruses (Goodman 1980; Goodman and Bird 1978; Rose 1978).

Geminiviruses are also of considerable fundamental biological interest. They are unique among all viruses in their virion morphology (paired or geminate icosahedra) (Fig. 6.1) and their unusually small size. The viral DNA is unusually small for an autonomous virus. Among eukaryotic viruses geminiviruses are unique in possessing covalently closed circular ssDNA. Since the discovery of these viruses in the mid-1970s and the

The work in my laboratory leading to the discovery of bean golden mosaic virus and the characterization of its genome as single-stranded DNA was supported by research contracts from the U.S. Agency for International Development to the University of Illinois International Soybean Program (INTSOY) and by the Illinois Agricultural Experiment Station. Current research and the writing of this chapter were supported by these agencies and by a grant (no. 7800549) from the USDA Competitive Research Grants Office.

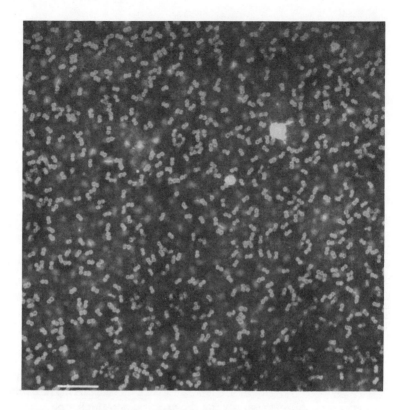

FIG. 6.1. Bean golden mosaic particles fixed in glutaraldehyde and negatively stained with sodium phosphotungstate, pH 7.0. Bar represents 200 nm.

realization that they contained circular ssDNA in 1977 (Goodman 1977a, 1977b; Harrison et al. 1977), much progress has been made in characterizing new geminiviruses (Table 6.1) and in defining their physical and chemical properties. Less well advanced is our knowledge about geminivirus replication and genetic organization.

It is knowledge about genetic functions and cell biology of host/virus interactions that will lead to possible ways to use geminiviruses in genetic engineering. Because of the newness of this field of investigation, it is not yet clear in a precise way if or how geminiviruses will fit into the future of genetic engineering with plants. The purpose of this brief chapter is to set forth what is presently known about the structure and function of geminiviruses and their unusual genomes. Whatever the utility of geminiviruses may turn out to be for genetic engineering, it seems clear even now that studies on the replication and pathogenicity of these unusual viruses will

TABLE 6.1

Geminiviruses and Their Properties

| Virus | Molecular Weights of Virions and Components | | | Vector* | References |
	DNA	Coat Protein	Virion		
Maize streak	7.1×10^5	2.8×10^4	n.a.	*Cicadulina mbila*	Harrison et al. 1977
Bean golden mosaic	8.0×10^5	2.74×10^4	3.9×10^6	*Bemisia tabaci*	Goodman et al. 1981
Cassava latent	8.0×10^5	3.4×10^4	n.a.	Unknown	Harrison et al. 1977
Chloris striate mosaic	7.1×10^5	2.79×10^4	3.8×10^6	*Nesoclutha pallida*	Francki et al. 1980
Tobacco yellow dwarf	n.a.	2.75×10^4	n.a.	*Orosius argentatus*	Thomas and Bowyer 1980

n.a. = not available

Bemisia tabaci is a whitefly; others are leafhoppers.

Note: In addition to the five viruses listed here, which have been at least partially characterized, at least five additional viruses are known that are probably members of the group. For details, see the review by Goodman (1981).

Source: Compiled by the author.

contribute to our understanding of genetic mechanisms in plants and, thereby, contribute indirectly, if not directly, to the future of genetic engineering for crop improvement.

PROPERTIES OF GEMINIVIRUSES

Most of the work on characterization of geminiviruses and their ssDNA has been done with bean golden mosaic virus (BGMV), chloris striate mosaic virus (CSMV), maize streak virus (MSV), and cassava latent virus (CLV). BGMV is a whitefly-transmitted virus originally isolated from *Phaseolus vulgaris* (Galvez and Castano 1976; Goodman, Bird, and Thongmeearkom 1977), in which it causes a serious, yield-limiting disease in parts of tropical America. CSMV and MSV are distinct leafhopper-transmitted viruses of grasses. CSMV was reported from Australia (Francki et al. 1979; Grylls 1963). MSV causes a major disease in maize in Africa (Bock, Guthrie, and Woods 1974). CLV was obtained from cassava with symptoms of cassava mosaic disease in Africa. The relationship between CLV and cassava mosaic disease is uncertain; CLV is sap transmitted from cassava to *Nicotiana clevelandii* and from *N. clevelandii* to *N. clevelandii* but not from *N. clevelandii* to cassava (Bock et al. 1977; Bock and Guthrie 1976; Bock, Guthrie, and Meredith 1978). The insect vector of CLV, if any, is unknown. The causal agent of cassava mosaic disease is transmitted by the same whitefly, *Bemisia tabaci*, that transmits BGMV (Storey and Nichols 1938). BGMV, like CLV, is sap transmissible. Neither MSV nor CSMV have been transmitted via sap.

All four viruses consist of paired particles with monomer units 18 to 20 nm in diameter (Fig. 6.1). Detailed studies on the structure of CSMV (Hatta and Francki 1979) indicate that the virus particles are constructed from two incomplete icosahedra with 22 capsomeres arranged in a $T = 1$ surface lattice. The exact arrangement of capsomeres at the interface of the two monomer units is not clear. With allowances for differences in preparation methods, the published pictures of all geminiviruses are very much alike, and it is therefore reasonable to conclude that their particles are organized in a similar manner. Published values for the molecular weight of the coat protein estimated by polyacrylamide gel electrophoresis are in the range of 27,000 to 34,000 (Table 6.1). Geminiviruses appear to have a single coat protein subunit with a tendency to dimerize. Preparations of highly purifed BGMV contain, in addition, two slightly smaller minor components, one of which coelectrophoreses with a major nuclear protein (Goodman et al. 1980); the possible roles of these additional proteins in virus structure are not understood.

BGMV, CSMV, MSV, and CLV have been shown to contain covalently closed circular ssDNA. The molecular weights of the DNAs are about 7×10^5 (MSV, CSMV) or 8×10^5 (BGMV, CLV) and are based on electron microscopic measurements in which the ssDNA of ϕX174 was used as a standard in each case (Fig. 6.2) (Francki et al 1980; Goodman 1977b; Harrison et al 1977; Reisman, Ricciardi, and Goodman 1979). There is no evidence for the covalent association of any other macromolecular constituents in the viral DNA. The DNAs are resistant to ribonuclease and alkali treatment, ruling out involvement of ribonucleotides in the structure (Francki et al. 1980; Goodman 1977a,1977b; Harrison et al. 1977). In the one case where covalently bound proteins were looked for, there was no evidence of such a constituent (Goodman et al. 1980).

Most geminivirus DNA preparations contain linear ssDNA as well as the circular molecules. The linear molecules from BGMV are nearly as long as the circular molecules (Reisman, Ricciardi, and Goodman 1979), whereas from the other viruses the majority of linear DNA molecules are much shorter (Francki et al. 1980; Harrison et al. 1977). Several lines of evidence indicate that the linear DNA is derived from breakage of the circular molecules. BGMV DNA made by a very gentle procedure sometimes contains no linear DNA. BGMV virus preparations contain trace amounts of an endonuclease that generates linear molecules during disruption of the virus with

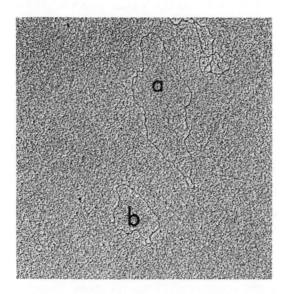

FIG. 6.2. DNA of bean golden mosaic virus (b) and ϕX174 (a). Preparation of DNAs and electron microscopy were as described by Reisman, Ricciardi, and Goodman (1979).

proteinase K (Goodman et al. 1980). In a careful analysis of the sedimentation properties of paired and monomer CSMV particles, a very high correlation was found between paired, presumably native, virus particles and the proportion of circular DNA molecules present (Francki et al. 1980).

Finally, for both CSMV and BGMV recent evidence shows that the proportion of DNA to protein in these geminiviruses is about 19 to 20 percent (Francki et al. 1980; Goodman et al. 1980). This result dictates that only one DNA molecule of $7–8 \times 10^5$ daltons is present in each *paired* particle. Since in routine BGMV DNA preparations 90 percent of the DNA molecules are circular, at least 90 percent of the virus particles contain circular DNA (Goodman et al. 1980). A recurring theme in early reports on viruses now included among the geminiviruses was the idea that the geminate configuration represented in a physical way the genetic requirement for two genome components (Bock, Guthrie, and Woods 1974; Galvez and Castano 1976). Results of physical measurements of these particles appear to rule out this hypothesis; the results do not, however, rule out the possibility that geminiviruses are multicomponent viruses but only that the geminate particles are not the physical manifestation of multicomponency (Francki et al. 1980; Goodman et al. 1980).

RELATIONSHIPS WITH OTHER ssDNA VIRUSES

On the basis of their physical properties, the geminiviruses are distinct from all other viruses. Among viruses that contain genomes of ssDNA, they are also unique. All prokaryotic ssDNA viruses, whether of icosahedral or helical symmetry or bacilliform, have circular ssDNA genomes with molecular weights in the range of $1.7–2.0 \times 10^6$. None of the bacterial or mycoplasma viruses have geminate virions (Table 6.2).

Other eukaryotic ssDNA viruses include the autonomous parvoviruses of mammals, the densonucleosis viruses of insects, and the defective adeno-associated viruses. All have linear ssDNA with molecular weights in the range of $1.5–2.0 \times 10^6$. Unlike the autonomous mammalian parvoviruses, the other animal ssDNA viruses package both complementary strands in different particles so that the virions contain ssDNA but the viral DNA obtained from phenol extraction and ethanol precipitation is predominantly double stranded (Rose et al. 1969). The parvoviruses, the only animal viruses known with ssDNA, are all isometric and none are geminate.

Geminiviruses are thus the smallest ssDNA viruses known and the only viruses of eukaryotes that contain covalently closed circular ssDNA. The size of the viral DNA molecule in geminiviruses is less than half that of any other ssDNA virus. There are no geminiviruses yet known in animals or bacteria.

TABLE 6.2

Summary of the Properties of Autonomous Single-Stranded DNA Viruses

Taxon of Host	Example	Virion Conformation	DNA		Reference
			MW ($\times 10^{-6}$)	Topology	
Prokaryotes					
Bacteria	φX174	Isometric	1.7	Circular	Sanger et al. 1978
Bacteria	fd	Filamentous	1.9	Circular	Berkowitz and Day 1974
Mycoplasma	MVL51	Bacilliform	1.5	Circular	Nowak, Maniloff, and Das 1978
Eukaryotes					
Mammals	MVM	Isometric	1.5	Linear	Bourguignon, Tattersall, and Ward 1976
Insects	DNV	Isometric	2.0	Linear	Kelly and Bud 1978
Plants	BGMV	Geminate	0.8	Circular	Reisman, Ricciardi, and Goodman 1979

Source: Compiled by the author.

CELL BIOLOGY AND REPLICATION OF GEMINIVIRUSES

Mature geminivirus particles accumulate and cause distinctive alterations in the nuclei of infected cells. In the most carefully studied cases, BGMV (Kim, Shock, and Goodman 1978), Euphorbia mosaic virus (EMV) (Kim and Flores 1979), and beet curly top virus (CTV) (Esau 1976; Esau and Hoeffert 1973; Esau and Magyarosy 1979), virus particles are seen only in phloem-associated cells. The whitefly-transmitted BGMV and EMV appear to cause similar major cytopathic effects; the nucleolar material becomes segregated into amorphous and granular regions and fibrillar rings containing deoxyribonucleoprotein are found. The fibrillar rings have been postulated but not proved to be the sites of viral DNA synthesis. Many of the nucleopathic effects of BGMV infection are grossly similar to those seen in the infection of cultured mammalian cells infected with the parvovirus H-1 (Kim, Shock, and Goodman 1978; Singer and Toolan 1975); the underlying basis for this similarity has not been investigated. Virus particles are seen in the cytoplasm only in mature, enucleate sieve elements, and no other characteristic features, such as inclusion bodies, are found in the cytoplasm of infected cells.

Beet CTV differs from the other geminiviruses studied at the ultra-structural level in causing cell hypertrophy and hyperplasia, in the depletion of nuclear chromatin that accompanies the appearance of virus particles, and in the absence of virus particles in sieve elements. CTV infection is also consistently associated with an uncharacterized amorphous inclusion body in the cytoplasm (Esau and Hoeffert 1973; Esau and Magyarosy 1979).

While confined to the phloem in systemically infected bean plants, BGMV appears to be able to infect other types of cells. Because BGMV is sap transmissible, we infer that the epidermal and probably mesophyll cells of young expanding primary leaves of the susceptible host plants can be infected. Further, recent work in my laboratory (Bajet and Goodman, unpubl.) shows that a high proportion of mesophyll protoplasts prepared from bean primary leaves becomes infected during a 5 minute inoculation treatment with BGMV DNA. The tissue restriction of BGMV in intact plants is apparently not because cells in other tissues are nonsusceptible.

The geminiviruses with monocotylendonous host ranges, MSV and CSMV, have not been sap transmitted. Cells of all tissues except the epidermis appear to become infected in systemic infections by MSV and CSMV (Bock, Guthrie, and Woods 1974; Francki et al. 1979). It is not known what the barrier to infection of the epidermis is. No detailed cytological study of MSV or CSMV has been published.

Little is known about the replication of geminiviruses, although some novel aspects of nucleic acid replication in plants can be inferred from the nature of the viral DNA and the unusual cytological features displayed by BGMV-infected cells. Our current thinking about the possible pathways of replication of BGMV are based in part on the model of replication of the only other circular ssDNA viruses known — those of bacteria and myco-plasmas — and in part on questions not yet answered conclusively about the genetic organization of BGMV.

To deal with the latter questions first, the circular DNA molecule in BGMV and other geminiviruses is remarkably small. Results mentioned earlier (Francki et al. 1980; Goodman et al. 1980) appear to rule out the linear DNA molecules as a necessary part of the BGMV or CSMV genomes. Our recent work with BGMV shows that the circular DNA of BGMV, puri-fied by electrophoresis in 4 percent polyacrylamide gels containing 7 M urea, is infective when inoculated to bean protoplasts. These experiments were done using an indirect fluorescent antibody method to detect viral an-tigen in DNA-inoculated and mock-inoculated control protoplasts. Unfrac-tionated BGMV DNA containing both circular and linear molecules and mixtures of purified circular and linear molecules are no more infective than the circular DNA alone. From these results and the evidence described above that circular DNA appears to be what is contained in intact geminate particles, we conclude that the complete genetic information of these viruses is contained in circular DNA molecules.

We next considered whether circular DNA molecules contain one se-quence or whether there might possibly be two or more circles of the same physical size but differing in nucleotide sequence and thus genetic content. Preliminary evidence that BGMV DNA consists of circular molecules con-taining more than one sequence comes from results of two types of experi-ments (Haber et al. 1981). Analysis of infectivity dilution curves using bean protoplasts and purified viral DNA gives a slope corresponding to two-hit rather than one-hit kinetics. BGMV DNA labeled in vivo with ^{32}P was digested with the restriction endonuclease HhaI, which has been shown to restrict ϕX174 ssDNA site-specifically with the same specificity (GCG$^{\prime}$C) as in dsDNA (Blakesley and Wells 1975). The resulting fragments were separated by electrophoresis and their sizes compared with ϕX174 ssDNA fragments generated by restriction with endonuclease HaeIII; the results showed that the BGMV fragments have a total molecular weight greater than the known 8×10^5 molecular weight of BGMV DNA (Reisman, Ric-ciardi, and Goodman 1979). Results of these experiments also showed that circular and linear BGMV DNA restricted with endonuclease HhaI gave es-sentially the same fragment pattern, adding additional evidence supporting the conclusion above that linear and circular DNA contain the same nucleo-tide sequences. Thus, it may be that the geminiviruses contain multipartite genomes, although not in the way that was originally expected (Bock, Guthrie, and Woods 1974; Francki et al. 1980; Galvez and Castano 1976).

BGMV-infected beans contain an as yet uncharacterized double-stranded DNA with sequences that hybridize with viral DNA (Ikegami and Goodman, unpubl.). In neutral sucrose density gradients, this virus-specific nucleic acid sediments slightly faster than viral ssDNA and on this basis is thought possibly to be of unit size in comparison with the size of viral DNA.

Geminivirus DNA replication probably depends at least in part on host cell machinery. This seems a reasonable speculation even if the genome turns out to be more complex than the physical size of the circular DNA molecules. The circular ssDNA bacterial virus ϕX174, whose replication strategy is known in some detail (Dressler et al. 1978), contains genes for three structural polypeptides (geminiviruses contain only one structural protein) and seven nonstructural polypeptides coded in a region of the genome containing overlapping genes (Barrell, Air, and Hutchinson 1976; Godson et al. 1978). Nevertheless, several critical steps in the replication cycle of the virus are accomplished by host enzymes, including priming of the synthesis of complementary (−) DNA, DNA synthesis by DNA polymerase III holoenzyme, DNA binding protein, and others (McMacken et al. 1978).

We do not know whether geminivirus DNA replication occurs via a mechanism similar to that of ϕX174. There are other possibilities. Incoming circular DNA molecules could be converted to a self-primed linear form by a site-specific cleavage, possibly by an enzyme carried in the virion (Goodman et al. 1980) or by a preexisting enzyme in the host cell. Subsequent DNA replication might then follow a mechanism similar to that of the autonomous parvoviruses (Berns and Hauswirth 1978; Ward and Tattersall 1978). Another alternative is RNA-mediated replication. RNA polymerase, which does not require a primer, might be used either to prime DNA replication, as for the bacterial virus M13 (McMacken et al. 1978), or possibly to synthesize a messenger RNA (mRNA) complementary to the incoming viral DNA coding for a primase or some other crucial first viral gene product necessary for replication of circular ssDNA. In either case, the means by which DNA polymerase activity on the viral template is primed is of central importance, as no DNA synthesizing enzyme is known that can initiate DNA synthesis on a DNA template without a primer. Yet another possibility is for viral DNA synthesis to be entirely RNA mediated. A DNA/RNA hybrid might be an intermediate, with progeny viral DNA synthesized by a reverse transcriptase; however, preliminary evidence for a dsDNA in infected tissues with sequences that hybridize with viral DNA favors a DNA-mediated rather than RNA-mediated mechanism.

PROSPECTS FOR GENETIC ENGINEERING

Much remains to be learned about the genetic organization and mode of replication of geminiviruses before the possible utility of these viruses in genetic engineering will be clear. Gene products in addition to the coat protein need to be identified. It will be important to know whether the viruses replicate through a DNA or an RNA intermediate and whether the double-stranded replication intermediate — which on the basis of preliminary evidence is dsDNA — is infective as is the case with the ssDNA bacterial viruses. Physical and transcription mapping of the dsDNA is needed to provide information on the location of essential virus gene functions so that other genetic information can be inserted without eliminating the ability of the viral DNA to enter, and replicate in, the cell. It will also be important to know what effect, if any, the virus has on DNA synthesis in host cells.

At this early stage of knowledge concerning geminiviruses and genetic mechanisms in plants, these viruses may be useful both as vehicles for introducing new genetic functions into higher plant cells and as vectors for cloning DNA in plant cells. Viral replication results in a large amplification of viral DNA, and it is not unreasonable to speculate that in the future this system could be used for cloning in plants as, for example, the circular ssDNA bacterial virus M13 that is used for cloning in prokaryotes (Messing et al. 1977; Ray and Kook 1978). Presumably, plant cell protoplasts could be used for this purpose, although many advances in the biochemical genetics of the viruses and the cell system will be necessary before such a possibility comes to fruition.

Also of interest is the possibility of using these small viruses as vehicles to introduce and study the expression of new genes in attempts to understand genetic mechanisms in higher plants. Geminiviruses may provide a useful complement to studies using the Ti-plasmid (see Chapter 8) and cauliflower mosaic virus (see Chapter 7). Geminiviruses replicate in the nucleus of infected cells, whereas cauliflower mosaic virus replication occurs in the cytoplasm (Shepherd 1979). The Ti-plasmid is in part integrated into the host DNA whereas there is as yet no evidence that geminivirus DNA is integrated, at least in the protoplast system where productive infection occurs.

The ability of geminiviruses to replicate in mesophyll cells opens the prospect of introducing and obtaining expression of nonviral genes on an engineered virus in the nucleus of plant cells. By well-established plant tissue culture techniques one might then obtain differentiated plant structures and whole plants with new phenotypes. The agricultural utility of this approach could be quite significant; many important crop species — including maize, wheat, soybeans, beans, tomatoes, sugar cane, and beets — are susceptible to one or more geminiviruses. Several geminiviruses have quite wide host ranges, and the use of protoplast technology may well allow use of the less promiscuous geminiviruses with plant species other than those that are susceptible as intact plants.

To the extent that small size is an advantage in contemplating recombinant DNA experiments, geminiviruses are clearly leading candidates, as the circular DNA genomes of these viruses are the smallest gene-encoding, autonomously replicating nucleic acids known. The problem of pathogenicity remains a challenge with these and other plant viruses. But perhaps it is with the geminiviruses that our best chance rests to separate recognition and replication functions from pathogenicity. In any event, essentially symptomless strains of numerous other plant viruses have been found, and it is thus probably only a matter of time and work before a variety of useful geminivirus mutants — including some with reduced pathogenicity — will be available for possible use in genetic engineering.

REFERENCES

Barrell, B. G.; Air, G. M.; and Hutchinson, C. A., III. 1976. Overlapping genes in bacteriophage φX174. *Nature* (London) 264:34–41.

Berkowitz, S. A., and Day, L. A. 1974. Molecular weight of single-stranded fd bacteriophage DNA. High speed equilibrium sedimentation and light scattering measurements. *Biochemistry* 13:4825–31.

Berns, K. I., and Hauswirth, W. W. 1978. Parvovirus DNA structure and replication. In *Replication of mammalian parvovirus*, ed. D. C. Ward and P. Tattersall, pp. 13–32. Cold Spring Harbor, N.Y.: Cold Spring Harbor Laboratory.

Blakesley, R. W., and Wells, R. D. 1975. "Single-stranded" DNA from φX174 and M13 is cleaved by certain restriction endonucleases. *Nature* (London) 257: 421–22.

Bock, K. R., and Guthrie, E. J. 1976. Recent advances in research on cassava viruses in East Africa. In *African cassava mosaic*, ed. B. L. Nestel, pp. 11–16. Ottawa, Canada: International Research and Development Centre.

Bock, K. R.; Guthrie, E. J., and Meredith, G. 1978. Distribution, host range, properties, and purification of cassava latent virus, a geminivirus. *Ann. Appl. Biol.* 90:361–67.

Bock, K. R.; Guthrie, E. J.; Meredith, G.; and Barker, H. 1977. RNA and protein components of maize streak and cassava latent viruses. *Ann. Appl. Biol.* 85:305–8.

Bock, K. R.; Guthrie, E. J.; and Woods, R. D. 1974. Purification of maize streak virus and its relationship to viruses associated with streak diseases of sugarcane and *Panicum maximum*. *Ann. Appl. Biol.* 77:289–96.

Bourguignon, G. J.; Tattersall, P. J.; and Ward, D. C. 1976. DNA of minute virus of mice: self-priming, nonpermuted, single-stranded genome with a 5'-terminal hairpin duplex. *J. Virol.* 20:290–306.

Dressler, D.; Hourcade, D.; Koths, K.; and Sims, J. 1978. The DNA replication cycle of the isometric phages. In *The single stranded DNA phages*, ed. D. T. Denhardt, D. Dressler, D. S. Ray, pp. 187–214. Cold Spring Harbor, N.Y.: Cold Spring Harbor Laboratory.

Esau, K. 1976. Hyperplastic phloem and its plastids in spinach infected with the curly top virus. *Ann. Bot.* 40:637–44.

Esau, K., and Hoeffert, L. L. 1973. Particles and associated inclusions in sugarbeet infected with the curly top virus. *Virology* 56:454-64.

Esau, K., and Magyarosy, A. C. 1979. Nuclear abnormalities and cytoplasmic inclusion in *Amsinckia* infected with the curly top virus. *J. Ultrastruct. Res.* 66:11-21.

Francki, R. I. B.; Hatta, T.; Boccardo, G.; and Randles, J. W. 1980. The composition of chloris striate mosaic virus, a geminivirus. *Virology* 101:233-41.

Francki, R. I. B.; Hatta, T.; Grylls, N. E.; and Grivell, C. J. 1979. The particle morphology and some other properties of chloris striate mosaic virus. *Ann. Appl. Biol.* 91:51-59.

Galvez, G. E., and Castano, M. 1976. Purification of the whitefly-transmitted bean golden mosaic virus. *Turrialba* 26:205-7.

Godson, G. N.; Fiddes, J. C.; Barrell, B. G.; and Sanger, F. 1978. Comparative DNA sequence analysis of the G4 and ϕX174 genomes. In *The single stranded DNA phages,* ed. D. T. Denhardt, D. Dressler, and D. S. Ray, pp. 51-86. Cold Spring Harbor, N.Y.: Cold Spring Harbor Laboratory.

Goodman, R. M. 1977a. Infectious DNA from a whitefly-transmitted virus of *Phaseolus vulgaris. Nature* (London) 266:54-55.

——. 1977b. Single-stranded DNA genome in whitefly-transmitted plant virus. *Virology* 83:171-79.

——. 1981. Geminiviruses. In *Handbook of plant virus diseases and comparative diagnosis,* ed. E. Kurstak. Amsterdam: Elsevier/North-Holland Biomedical Press.

Goodman, R. M., and Bird, J. 1978. Bean golden mosaic virus. In *Descriptions of plant viruses no. 192.* Kew, England: Commonwealth Mycological Institute.

Goodman, R. M.; Bird, J.; and Thongemeearkom, P. 1977. An unusual virus-like particle associated with golden yellow mosaic of beans. *Phytopathology* 67: 37-42.

Goodman, R. M.; Shock, T. L.; Haber, S. M.; Browning, K. S.; and Bowers, G. R., Jr. 1980. The composition of bean golden mosaic virus and its single-stranded DNA genome. *Virology* 106:168-72.

Grylls, N. E. 1963. A striate mosaic virus disease of grasses and cereals in Australia, transmitted by the cicadellid *Nesoclutha obscura. Austr. J. Agric. Res.* 14: 143-53.

Haber, S.; Ikegami, M.; Bajet, N. B.; and Goodman, R. M. 1981. Evidence for a divided genome in bean golden mosaic virus, a geminivirus. *Nature* (London) 289:324-26.

Harrison, B. D.; Barker, H.; Bock, K. R.; Guthrie, E. J.; Meredith, G.; and Atkinson, M. 1977. Plant viruses with circular single-stranded DNA. *Nature* (London) 270:760-62.

Hatta, T., and Francki, R. I. B. 1979. The fine structure of chloris striate mosaic virus. *Virology* 92:428-35.

Kelly, D. C., and Bud, H. M. 1978. Densonucleosis virus DNA: analysis of fine structure by electron microscopy and agarose gel electrophoresis. *J. Gen. Virol.* 40:33-43.

Kim, K. S., and Flores, E. M. 1979. Nuclear changes associated with *Euphorbia* mosaic virus transmitted by the whitefly. *Phytopathology* 69:980-84.

Kim, K. S.; Shock, T. L.; and Goodman, R. M. 1978. Infection of *Phaseolus vulgaris* by bean golden mosaic virus: ultrastructural aspects. *Virology* 89:22–33.

McMacken, R.; Rowan, L.; Ueda, K.; and Kornberg, A. 1978. Priming of DNA synthesis on viral single-stranded DNA in vitro. In *The single stranded DNA phages*, ed. D. T. Denhardt, D. Dressler, and D. S. Ray, pp. 273–86. Cold Spring Harbor, N.Y.: Cold Spring Harbor Laboratory.

Matthews, R.E.F. 1979. *Plant Virology*. New York/London: Academic Press.

Messing, J.; Gronenborn, B.; Muller-Hill, B.; and Hofschneider, P. H. 1977. Filamentous coliphage M13 as a cloning vehicle: insertion of a *Hind*II fragment of the *lac* regulatory region in M13 replicative form in vitro. *Proc. Nat. Acad. Sci. (USA)* 74:3642–46.

Nowak, J.; Maniloff, J.; and Das, J. 1978. Electron microscopy of single-stranded mycoplasmavirus DNA. *Fed. Evr. Microbiol. Soc. Lett.* 4:59–61.

Ray, D. S., and Kook, K. 1978. Development of M13 as a single-stranded cloning vector: insertion of the Tn3 transposon into the genome of M13. In *The single stranded DNA phages*, ed. D. T. Denhardt, D. Dressler, and D. S. Ray, pp. 455–60. Cold Spring Harbor, N.Y.: Cold Spring Harbor Laboratory.

Reisman, D.; Ricciardi, R. P.; and Goodman, R. M. 1979. The size and topology of single-stranded DNA from bean golden mosaic virus. *Virology* 97:388–95.

Rose, D. J. W. 1978. Epidemiology of maize streak disease. *Ann. Rev. Entomol. 23: 259–82.*

Rose, J. A.; Berns, K. I.; Hoggan, M. D.; and Koczot, F. J. 1969. Evidence for a single-stranded adenovirus-associated virus genome. Formation of a DNA density hybrid on release of viral DNA. *Proc. Nat. Acad. Sci. (USA)* 64: 863–69.

Sanger, F.; Coulson, A. R.; Friedmann, T.; Air, G. M.; Barrell, B. G.; Brown, N. L.; Fiddes, J. C.; Hutchinson, C. A., III; Slocombe, P. M.; and Smith, M. 1978. The nucleotide sequence of bacteriophage ϕX174. *J. Mol. Biol.* 125:225–46.

Shepherd, R. J. 1979. DNA plant viruses. *Ann. Rev. Plant Physiol.* 30:405–23.

Singer, I. I., and Toolan, H. 1975. Ultrastructural study of H-1 parvovirus replication. I. Cytopathology produced human NB epithelial cells and hamster embryo fibroblasts. *Virology* 65:40–54.

Storey, H. H., and Nichols, R. F. W. 1938. Studies on the mosaic diseases of cassava. *Ann. Appl. Biol.* 25:790–806.

Thomas, J. E., and Bowyer, J. W. 1980. Properties of tobacco yellow dwarf and bean summer death viruses. *Phytopathology* 70:214–17.

Ward, D. C., and Tattersall, P., eds. 1978. *Replication of mammalian parvoviruses*. Cold Spring Harbor, N.Y.: Cold Spring Harbor Laboratory.

7

CAULIFLOWER MOSAIC VIRUS DNA AS A POSSIBLE GENE VECTOR FOR HIGHER PLANTS

Roger Hull

INTRODUCTION

The classical methods of plant breeding — the use of genetic recombination and selection — have some limitations, the main one being mating incompatibility. It is, therefore, not always possible to introduce desired genes. Genetic engineering techniques, which involve the direct transfer of genes into the cell, can theoretically overcome this barrier. Work over recent years (for review, see Kleinhofs and Behki 1977) has shown that although plant cells can take up pieces of deoxyribonucleic acid (DNA), these do not (with the exception of viral DNA) replicate and express. By analogy with prokaryotic systems, it would appear that an introduced gene needs to have a "vector" that would control its replication and expression. Among the possible vectors in plant systems is the DNA of double-stranded DNA plant viruses. There is just such a group of plant viruses, the caulimoviruses.

As well as being potential gene vectors, the caulimoviruses can also be used as a model system for studying DNA replication and expression in plants. At present little is known about these functions at the molecular biological level.

Much of the work so far on caulimoviruses has been directed to the characterization of the viruses and their DNA. The study of the replication and expression of the DNA is just starting. In this chapter I will review the present state of knowledge on caulimoviruses and try to indicate some of the problems that will need to be overcome in examining their potential as gene vectors.

THE CAULIMOVIRUSES

The caulimovirus group comprises six viruses (carnation etched ring, cauli-flower mosaic (CaMV), dahlia mosaic, figwort mosaic, *Mirabilis* mosaic, and strawberry vein-banding viruses); cassava vein mosaic and petunia vein-clearing viruses may also be members of this group. Each of these viruses has a limited host range, and each systemically infects most of its hosts, commonly causing vein-banding mosaics. Most of the caulimoviruses are transmitted by aphids, the relationship with the vector being passive — that is, they do not multiply within the vector. However, for at least CaMV, a transmission factor is required for aphid transmission (Lung and Pirone 1973). The properties of members of the caulimovirus group have been reviewed by Shepherd (1976;1978;1979).

The most studied of the caulimoviruses is CaMV, and the rest of this chapter will be devoted to this virus.

CAULIFLOWER MOSAIC VIRUS

The particles of CaMV sediment at about 210s and are isometric and about 50 nm in diameter. They are composed of protein and DNA. Electron microscopy does not reveal any surface structure; however, a recent neutron diffraction study (Chauvin et al. 1979) has shown that the DNA is in direct contact with the protein shell and that the center of the virion appears to be free of DNA and protein. The protein shell was considered to be composed of several polypeptide species (Brunt et al. 1975; Kelly, Cooper, and Walkey 1974; Hull and Shepherd 1976; Tezuka and Taniguchi 1972), but it has recently been suggested (Al Ani, Pfeiffer, and Lebeurier 1979) that these arise from partial proteolysis and/or partial aggregation of a single species of molecular weight 42,000.

In the cell most of the virus particles are found embedded in inclusion bodies (Fig. 7.1) of apparently amorphous material. The inclusion bodies are found only in the cell cytoplasm, and at least for CaMV, very few particles are observed outside these bodies. Shepherd and Wakeman (1977) have shown that the inclusion bodies are comprised mainly of a protein of 55,000 molecular weight; however, recent estimates (Covey, unpubl.) indicate that the molecular weight is closer to 60,000.

THE DNA

The inclusion bodies have to be disrupted to purify CaMV particles. This is best effected by treatment of sap extracts with Triton X-100 and urea (Hull,

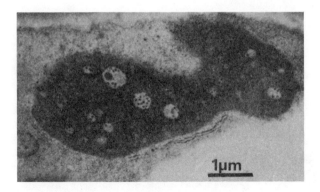

FIG. 7.1. Electron micrograph of thin section of CaMV-infected turnip leaf showing an inclusion body containing virus particles.

Shepherd, and Harvey 1976). Similarly, it is somewhat difficult to extract the DNA from virus particles; one has to digest the protein with proteases (Shepherd, Bruening, and Wakeman 1970). The DNA is double stranded, open-circular, with a molecular weight of about 5×10^6 (7,500–8,000 base pairs); the properties of the DNA are reviewed by Hull (1979). Some of the properties are very unusual. A marked proportion of the molecules have a twisted structure when viewed in the electron microscope (Fig. 7.2); this structure is also revealed in gel electrophoresis (see Hull 1979). The twisted-ness is not relaxed when the DNA is cut by a restriction endonuclease at a single site (Fig. 7.2) but is partially relaxed upon treatment with the protease papain (Hull and Wells, unpubl.).

There have been several restriction endonuclease maps of CaMV DNA published (Hull 1980b; Hull and Howell 1978; Lebeurier et al. 1978; Meagher, Shepherd, and Boyer 1977; Volovitch, Drugeon, and Yot 1978; Volovitch et al. 1979). There is, on the whole, broad agreement between these maps, but there are differences between the DNAs of different CaMV isolates. There are also some differences in presentation of maps, which could be confusing to readers. The current map for our standard isolate, Cabb B-JI is shown in Fig. 7.3. This differs from the map published by Hull (1980b) in that sites for AvaI and HpaI have been found and also that a further EcoRI site (0.545 map units) has been recognized. This last site gives a fragment of 62 base pairs, which has only been resolved using 10 percent polyacrylamide gels (Fig. 7.4, lane **B**).

Hull and Howell (1978) and Hull (1980b) have shown that there are differences between the DNAs of CaMV isolates. Thus far 33 isolates have been compared. Of these isolates 31 differ only in the presence or absence of a small number of sites when compared with the type isolate. One isolate,

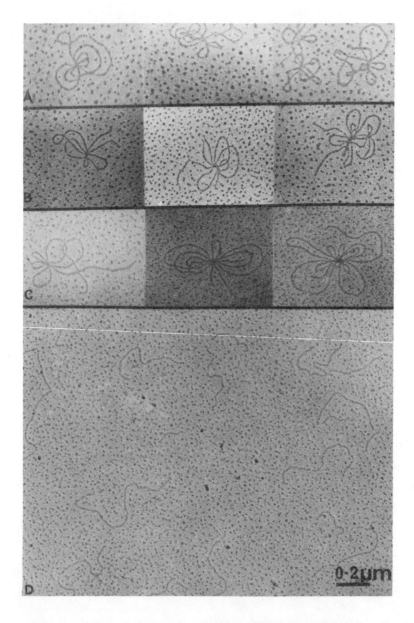

FIG 7.2. Electron micrographs of CaMV DNA. (A) Untreated molecules showing twisted structure. (B) Molecules cut with restriction endonuclease *Xho*I. (C) Molecules cut with restriction endonuclease *Sal*GI. (D) Molecules cut with restriction enzyme *Eco*RI.

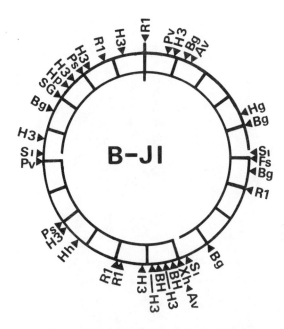

FIG. 7.3. Restriction endonuclease map of DNA of CaMV isolate Cabb B-JI.
Av = *Ava*I; **Bg** = *Bgl*II; **BH** = *Bam*HI; **Fs** = *Fsp*AI; **H3** = *Hind*III; **Hg** = *HgiAI*;
Hh = *Hha*I; **Hp** = *Hpa*I; **Ps** = *Pst*I; **Pv** = *Pvu*II; **RI** = *Eco*RI; **S**$_I$ = S$_I$ single-
strand nuclease; **SG** = *Sal*GI; **Xh** = *Xho*I.

CM4-184, has a deletion covering the region between map units 0.44–0.49
on the type DNA (Fig. 7.3); this isolate has lost the sites for S$_I$, *Xho*I,
*Bam*HI, *Hind*III, *Bam*HI, and *Hind*III. Hull and Howell (1978) and Hull
(1980b) reported an isolate that appeared to have a small addition in this
region. However, analyses of digests of these DNAs on 10 percent polyacrylamide gels (Fig. 7.4, lane **A**) have shown a lack of the 62 base pair fragment. Therefore, the apparent addition is really the loss of the *Eco*RI site at
0.54 map units. Similarly, the apparent deletion between map units 0.94
and 0.00 in the DNA of the Milan isolate has been shown to be due to an extra *Eco*RI site, probably at 0.995 map units. The other isolate whose restriction endonuclease map differs markedly from that of Cabb B-JI DNA is Bari
I. Comparison of the restriction endonuclease map of Bari I DNA (Fig. 7.5)
with that of the type isolate (Fig. 7.3) shows that Bari DNA has only 9 sites
(2 *Eco*RI, 2 *Bgl*II, and 5 *Hind*III) in common with the 36 sites mapped on
Cabb B-JI DNA. Two of the other sites are also found in one or more of the
other isolates examined. Thus 20 of the sites are, at present, unique to Bari I.
    Another unusual feature of CaMV DNA is the presence of single-
stranded regions (nicks or gaps) at specific sites (Hull and Howell 1978;

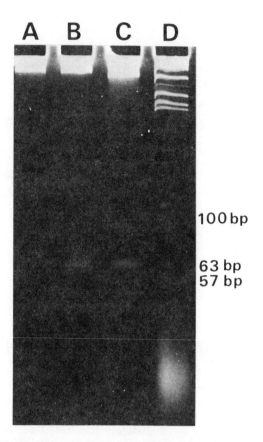

FIG. 7.4. Electrophoresis in 10 percent polyacrylamide gel of CaMV DNA cut by *Eco*RI. Lane **A**: Australian isolate; **B**: Cabb-B-JI; **C**: Milan; **D**: markers of pBR322 DNA cut by *Alu*I.

Volovitch, Drugeon, and Yot 1978; Volovitch et al. 1976). The majority of isolates have one gap in one strand and two in the other (Figs. 7.3 and 7.5). However, CM4-184, which has a deletion, has only one gap in each strand. Infill experiments using T₄ DNA polymerase indicate that these are truly gaps (not nicks) and that the total incorporation per DNA molecule is about 40 nucleotides (that is, about 12 nucleotides per gap) (Hull, unpubl.); it is not known if each gap is the same size. Hull et al. (1979) used the gaps to determine the polarity of CaMV DNA. They showed that the counterclockwise side of the gap at 0.24 map units and the clockwise side of the other two gaps were 5'. The 5' deoxyribonucleotides from each gap were isolated, and it was shown that those at 0.24 and 0.76 map units were pA and that at 0.44 map units was pG. This indicates that at least the 5' edges of the gaps

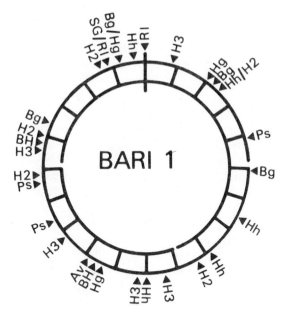

FIG. 7.5. Restriction endonuclease map of DNA of CaMV isolate Bari I.
Av = *Av*aI; **Bg** = *Bgl*II; **BH** = *Bam*HI; **H2** = *Hinc*II; **H3** = *Hind*III; **Hg** = *Hgi*AI;
Hh = *Hha*I; **Ps** = *Pst*I; **RI** = *Eco*RI; **SG** = *Sal*GI.

are in fixed positions. The gaps are not present in CaMV DNA that has been
cloned in *Escherichia coli* (Hohn et al. 1980). The reason for these gaps is at
present unknown. However, the fact that at least two of them are always
found in CaMV DNA from eukaryote sources but not in DNA from prokaryotic sources may reflect a major functional difference between eukaryotes and prokaryotes.

RIBONUCLEIC ACID TRANSCRIPTS

Ribonucleic acid (RNA) transcripts of CaMV DNA have been isolated both
from infected protoplasts (Howell and Hull 1978) and from infected plants
(Howell, Odell, and Dudley 1980; Hull et al. 1979). Only one strand of
CaMV DNA, that with the single gap, is transcribed. One of the products of
transcription is a large polyadenylated RNA, which has been estimated to
represent 50 to 75 percent (Howell and Hull 1978), or more recently about
85 percent (Hull, unpubl.), of the coding capacity of that strand of DNA.
Hull et al. (1979) used the polyadenylated RNAs to confirm their determination of the polarity of the DNA. This study also indicated that transcription

was initiated between map units 0.24 and 0.30 and was terminated between map units 0.00 and 0.24.

Smaller RNA transcripts of CaMV DNA have also been found. Howell, Odell, and Dudley (1980) reported on two CaMV-specific RNA species that comigrate with the 25s and 18s ribosomal RNAs in gels. The polyadenylated fraction contains predominantly the 18s RNA and can be translated to give a protein (molecular weight 66,000) that is thought to be inclusion body protein. The nonpolyadenylated fraction contains mainly the 25s species. Using the "northern" transfer techniques (Alwine, Kemp, and Stark 1977) in total RNA and isolating specific RNAs by hybridizing to CaMV DNA immobilized on diazotized cellulose paper (Alwine, Kemp, and Stark 1977), we have recognized up to five CaMV-specific RNAs (Hull and Covey, unpubl.). However, these experiments are complicated by a high background of apparently partially degraded CaMV RNA, even though ribosomal and other RNAs are undegraded. It is possible that the CaMV RNAs are turned over relatively rapidly. This problem may be overcome by a recent observation that CaMV-specific RNAs can be isolated from partially purified inclusion bodies (Hull, unpubl.).

DISCUSSION

It is obvious from this survey of the knowledge on CaMV and its DNA that we do not know yet whether it can be used as a gene vector. There are many questions that remain to be answered. We do not know if there are sites in CaMV DNA into which foreign DNA can be inserted. The mapping of restriction endonuclease sites in numerous CaMV isolates has revealed at least 10 enzyme/isolate combinations that give single cuts (Hull 1980a; 1980b); these are spread around the CaMV genome. If we can locate sites for insertion, we may also be able to determine the minimum amount of CaMV DNA needed as a vector. There are indications that at least some of CaMV DNA can be removed and the virus still function; CM4-184 has a deletion of about 5 percent and still functions well. The aphid transmission factor (Lung and Pirone 1973) is probably virus coded and is not essential in a genetic manipulation situation – in fact, it is a disadvantage.

It is not known if a vector/gene combination would need to be packaged as, say, a viruslike particle for it to spread through a plant. If it does need to be packaged, a maximum packaging size would need to be determined. As noted above, neutron scattering (Chauvin et al. 1979) indicates space in the center of the particle; however, there may be steric problems that prevent this being filled with DNA. The stability of vector/gene combinations is another unknown factor as is the control of expression in the cytoplasm. There is no evidence for CaMV DNA integrating into the host genome (Hull 1980a). However, a nonintegrating system has been demonstrated recently for animal cells (Mulligan, Howard, and Berg 1979).

Even if these problems cannot be overcome and it proves impossible to use CaMV DNA as a gene vehicle, the study of this DNA will advance our knowledge of how plant cells function.

ADDENDUM

Since this chapter was written, there have been several important papers published on CaMV. These are reviewed briefly below.

The complete nucleotide sequence of the DNA of the Strasbourg isolate (named in the paper as Cabb B-S isolate but not derived from the original Cabb B isolate) was published by Franck et al. (1980). This paper showed that the DNA comprised 8,024 base pairs and that at least two of the single-strand discontinuities (gaps) are, in fact, overlaps of several nucleotides. An analysis of open reading frames showed that there were six long open-reading frames in the α strand; five of the open reading frames overlapped the adjacent one to a certain extent. The authors suggested that coding region IV was the cistron for a coat protein precursor. The major noncoding region of the α strand spans the discontinuity in that strand (map units 91-4, taking the discontinuity as map unit 0).

Several studies (Delseny and Hull, unpubl.; Howell, Walker, and Dudley 1980; Lebeurier et al. 1980) have shown that CaMV DNA cloned at various sites in *E. coli* plasmids pBR322 or pAT153 and the bacteriophage λgt*WES* is infectious provided that it is cut out from the vector DNA; infectivity is lost if the vector DNA is retained or if deletions are made in the CaMV DNA. Cloned CaMV DNA regains the gaps (discontinuities) and the twisted structure (Fig. 7.2) on passage through turnip plants.

Several transcripts of CaMV DNA have been recognized in RNA extracts from CaMV-infected leaves (Covey and Hull, in press; Odell and Howell 1980). Two species, about 7-8 kb and 2.3 kb, accumulate radioactivity when CaMV-infected leaves were labeled with [^{32}P]-orthophosphate 20 days after inoculation. The 2.3 kb RNA is the messenger RNA for a 62,000 molecular weight polypeptide that was shown to be very similar to the major protein component of the virus inclusion bodies. This RNA is transcribed from the contiguous *Eco*RI fragments d and b (map units 0.94-0.30 in Fig. 7.3) — a region that corresponds to open reading frame VI as determined from the nucleotide sequence (Franck et al. 1980). There is some evidence for temporal changes in the relative amounts of these and other RNA transcripts.

REFERENCES

Al Ani, R.; Pfeiffer, P.; and Lebeurier, G. 1979. The structure of cauliflower mosaic virus. II. Identity and location of the viral polypeptides. *Virology* 93:188–97.

Alwine, J. C.; Kemp, D. J.; and Stark, G. R. 1977. Method for detection of specific RNAs in agarose gels by transfer to diazobenzyloxymethyl-paper and hybridization with DNA probes. *Proc. Nat. Acad. Sci. (USA)*. 74:5350–54.

Brunt, A. A.; Barton, R. J.; Tremaine, J. H.; and Stace-Smith, R. 1975. The composition of cauliflower mosaic virus protein. *J. Gen. Virol.* 27:101–6.

Chauvin, C.; Jacrot, B.; Lebeurier, G.; and Hirth, L. 1979. The structure of cauliflower mosaic virus: a neutron diffraction study. *Virology* 96:640–41.

Covey, S. N., and Hull, R. In press. Transcription of cauliflower mosaic virus DNA. Detection of transcripts, properties and isolation of the gene encoding the virus inclusion body protein. *Virology*.

Franck, A.; Guilley, H.; Jonard, G.; Richards, K.; and Hirth, L. 1980. Nucleotide sequence of cauliflower mosaic virus DNA. *Cell* 21:285–294.

Hohn, T.; Hohn, B.; Lebeurier, G.; and Lesot, A. 1980. Restriction map of original and cloned cauliflower mosaic virus DNA. *Gene* 11:21–31.

Howell, S. H., and Hull, R. 1978. Replication of cauliflower mosaic virus and transcription of its genome in turnip leaf protoplasts. *Virology* 86:468–81.

Howell, S. H.; Odell, J. T.; and Dudley, K. R. 1980. Expression of the cauliflower mosaic virus genome in turnips (*Brassica rapa*). In *Genome organization and expression in plants*, ed. C. E. Leaver, New York: Plenum Press.

Howell, S. H.; Walker, L. L.; and Dudley, R. K. 1980. Cloned cauliflower mosaic virus DNA infects turnips (*Brassica rapa*). *Science* 208:1265–67.

Hull, R. 1979. The DNA of plant DNA viruses. In *Nucleic acids in plants*, ed. J. W. Davies and T. C. Hall, vol. 2, pp. 3–29. West Palm Beach, Fla.: CRC Press.

Hull, R. 1980a. Genetic engineering in plants: the possible use of cauliflower mosaic virus DNA as a vector. In *Proceedings of the international workshop in plant cell cultures: results and perspectives*, ed. F. Sala, B. Pamsi, R. Cella, and O. Ciferri, pp. 219–24. Amsterdam: Elsevier/North-Holland.

——— . 1980b. Structure of the cauliflower mosaic virus genome. III. Restriction endonuclease mapping of thirty-three isolates. *Virology* 100:76–90.

Hull, R.; Covey, S. N.; Stanley, J.; and Davies, J. W. 1979. The polarity of the cauliflower mosaic virus genome. *Nucl. Acids Res.* 7:669–77.

Hull, R., and Howell, S. H. 1978. Structure of the cauliflower mosaic virus genome. II. Variation in DNA structure and sequence between isolates. *Virology* 86:482–93.

Hull, R., and Shepherd, R. J. 1976. The coat proteins of cauliflower mosaic virus. *Virology* 70:217–20.

Hull, R.; Shepherd, R. J.; and Harvey, J. D. 1976. Cauliflower mosaic virus: an improved purification procedure and some properties of the virus particles. J. Gen. Virol. 31:93–100.

Kelly, D. C.; Cooper, J.; and Walkey, D. G. A. 1974. Cauliflower mosaic virus structural proteins. *Microbios* 10:239–45.

Kleinhofs, A., and Behki, R. 1977. Prospects for plant genome modification by nonconventional methods. *Ann. Rev. Genet.* 11:79–101.

Lebeurier, G.; Hirth, L.; Hohn, T.; and Hohn, B. 1980. Infectivities of native and cloned DNA of cauliflower mosaic virus. *Gene* 12:139–46.

Lebeurier, G.; Whitechurch, O.; Lesot, A.; and Hirth, L. 1978. Physical map of DNA from a new cauliflower mosaic virus strain. *Gene* 4:213–26.

Lung, M. C. Y., and Pirone, T. P. 1973. Studies on the reason for differential transmissibility of cauliflower mosaic virus isolates by aphids. *Phytopathology* 63: 910–14.

Meagher, R. B.; Shepherd, R. J.; and Boyer, H. W. 1977. The structure of cauliflower mosaic virus. I. A restriction map of cauliflower mosaic virus DNA. *Virology* 80:362–75.

Mulligan, R. C.; Howard, B. H.; and Berg, P. 1979. Synthesis of rabbit β-globulin. *Nature* 227:108–14.

Odell, J. T., and Howell, S. H. 1980. The identification, mapping and characterization of mRNA for P66, a cauliflower mosaic virus-coded protein. *Virology* 102:349–59.

Shepherd, R. J. 1976. DNA viruses of higher plants. *Adv. Virus Res.* 20:305–39.

——. 1978. Cauliflower mosaic virus (DNA virus of higher plants). In *Atlas of insect and plant viruses*, ed. K. Maramarosch, pp. 159–66. New York: Academic Press.

——. 1979. DNA plant viruses. *Ann. Rev. Plant Physiol.* 30:405–23.

Shepherd, R. J.; Bruening, G. E.: and Wakeman, R. J. 1970. Double-stranded DNA from cauliflower mosaic virus. *Virology* 41:339–47.

Shepherd, R. J., and Wakeman, R. J. 1977. Isolation and properties of the inclusion bodies of cauliflower mosaic virus. *Proc. Amer. Phytop. Soc.* 4:145.

Tezuka, N., and Taniguchi, T. 1972. Structural protein of cauliflower mosaic virus. *Virology* 48:297–99.

Volovitch, M.; Drugeon, G.; Dumas, J. P.; Haenni, A. L.; and Yot, P. 1979. A restriction map of cauliflower mosaic virus DNA (strain PV 147). *Eur. J. Biochem.* 100:245–55.

Volovitch, M.; Drugeon, G.; and Yot, P. 1978. Studies on the single-stranded discontinuities of the cauliflower mosaic virus genome. *Nucl. Acids Res.* 5:2913–25.

Volovitch, M.; Dumas, J. P.; Drugeon, G.; and Yot, P. 1976. Single-stranded interruptions in cauliflower mosaic virus. In *Acides nucleiques et syntheses chez les vegetaux*, ed. L. Bogorad and J. H. Weil, pp. 635–41. Paris: Centre National de la Recherche Scientifique.

8

AGROBACTERIUM:
NATURE'S GENETIC ENGINEER

Robert C. Nutter
Michael F. Thomashow
Milton P. Gordon
Eugene W. Nester

INTRODUCTION

Until now plant characteristics have been altered by utilizing the classical methods of selective breeding or natural selection. With the discovery of restriction enzymes and the advent of molecular cloning techniques, today's molecular biologist has available powerful new tools with the potential to change the content or expression of genetic information in cells by "transplanting" genes, singly or in groups, to other desired locations or cellular backgrounds. In order to accomplish this in plants several factors must be considered: (1) the technology of cloning plant or other desirable genes, (2) the development of a suitable vehicle for introducing them into plant cells (and in the appropriate portion of the plant genome), and (3) understanding the mechanism(s) by which the expression of genetic information is normally controlled in plants.

In this chapter we will describe work that we have undertaken in Seattle to elucidate how a large plasmid (termed *Ti-plasmid*) of the soil bacterium *Agrobacterium tumefaciens* causes the plant disease crown gall. We feel that an understanding of the molecular biology of this system will allow the future exploitation of this natural genetic engineer to introduce genetic information into plant genomes. As the molecular biology of the Ti-plasmid and normal plant DNA becomes better understood, perhaps the second and third conditions listed above will be better met.

111

HISTORY OF CROWN GALL AND THE TI-PLASMID

It has been known for over 70 years that crown gall disease is caused by the gram-negative soil bacterium *A. tumefaciens* Conn (Smith and Townsend 1907). This pathogen has been found all over the world and has been shown to affect a wide variety of dicotyledonous plants (for reviews, see Gordon, forthcoming; Lippincott and Lippincott 1975). It is necessary for the plant to be wounded and for virulent actively metabolizing agrobacteria to be present in the wound. How virulence is conferred to the bacteria has only recently been elucidated.

Braun (1947) first postulated that a substance he termed the *tumor inducing principle* (TIP) was the active agent in tumorigenesis. However, it was not until 1974 that the observation was made that only virulent strains of agrobacteria harbored large (100-150 $\times$ 10^6 dalton) plasmids (Zaenen et al. 1974). This was followed by two reports. The first showed that when a virulent strain was "cured" of its large plasmid, it became avirulent (Van Larebeke et al. 1974). The second showed that the plasmid did not become integrated into the bacterial chromosome upon curing and that when the plasmid was transferred to an avirulent strain, this strain became virulent (Watson et al. 1975).

Two years later it was shown that axenically grown crown gall tumor cells had stably acquired a portion of the Ti-plasmid (Chilton et al. 1977). This firmly established that the Ti-plasmid was the TIP postulated by Braun and that the transfer of a portion of this plasmid to the plant was associated with the acquisition of the tumorous traits. Subsequent reports that the transferred plasmid deoxyribonucleic acid (termed *T-DNA*) was transcribed (Drummond et al. 1977; Gurley et al. 1979) and that tumor polysomal poly(A)$^+$ ribonucleic acid (RNA) released from T-DNA bound to nitrocellulose filters directed the in vitro synthesis of discrete polypeptides (McPherson, Nester, and Gordon 1980) have strengthened the findings of Chilton et al. (1977).

Merlo et al. (1980) extended the findings of Chilton et al. (1977) by determining the T-DNA complement in tumors incited by related plasmids. The three cloned tobacco tumor lines investigated were A6S/2, incited by pTi-A6; B6806/E9, incited by pTi-B$_6$806; and 15955/1, incited by pTi-15955. Isolated restriction fragments of the pTi-B$_6$806 virulence plasmid were radiolabeled with ^{32}P by nick translation and used as probes in solution hybridization reactions with various tumor DNAs driving the reactions. Analysis of the kinetics of renaturation showed that the extent of the T-DNA in the three tumor lines varied. B6806/E9 had more of the plasmid than 15955/1 or A6S/2. In addition, in the B6806/E9 line it was clear that the T-DNA complement was not represented equally across the entire region. Sequences representing the rightward region were present in multiple copies, whereas the sequences from the leftward region were present at only one to two copies. Analysis of 80 percent of the Ti-plasmid did not show any other regions of the plasmid present in the tumors. The limits of the technique and the inability to isolate adequate amounts of pure restriction fragments for radiolabeling prevented precise definition of the extent or organization of the T-DNA.

Two major types of Ti-plasmids have been identified and classified according to the ability of *A. tumefaciens* strains to utilize the rare amino acid derivatives octopine and nopaline as their sole source of carbon and nitrogen. In addition, these plasmids promote tumors that may synthesize either octopine or nopaline, respectively (Bomhoff et al. 1976). It has been shown that the octopine and nopaline plasmids share regions of DNA homology (Chilton, Drummond, Merlo, and Sciaky 1978; DePicker, Van Montagu, and Schell 1978) and that one particular region, termed the *common DNA*, might be important in determining virulence. There are other functions that have been ascribed to the Ti-plasmid, and the reader is referred to one of the numerous reviews for a listing.

The fact that a prokaryote is able to transfer a portion of its DNA to a eukaryote—and in effect genetically engineer a higher organism—has opened the possibility that scientists might also utilize the Ti-plasmid for the introduction of selected genes into plants of their choosing. A complete knowledge of the molecular biology of the Ti-plasmid, in addition to providing an understanding of cell transformation, will also hopefully lead to the development of a cloning vehicle suitable for plants. We have therefore attempted to answer basic questions regarding the molecular biology of crown gall disease. The three questions to be addressed in this chapter are:

What is the extent of the T-DNA in different tumor lines?
How is theT-DNA organized in the plant cell?
Is the T-DNA integrated into the plant genome?

CLONING OF THE TI-PLASMID

The development of molecular cloning techniques and the transfer of DNA in situ to nitrocellulose filters (Southern 1975) have provided powerful tools for the study of DNA sequences and their organization. We have cloned restriction fragments of the entire pTi-A6 plasmid into the *Escherichia coli* plasmid vehicles pBR322 or pBR325. Analysis of the homology of the cloned fragments resulting from double and triple digestion with appropriate restriction enzymes has permitted the constructing of a detailed map of the octopine Ti-plasmids in the region of the T-DNA. This is shown in Fig.

8.1. We have found no differences in the physical maps nor the amount of homology between pTi-A6, pTi-B₆806, or pTi-15955 in this region.

The use of Ti-plasmid cloned fragments has a number of advantages. First of all, they are not contaminated by other Ti-plasmid fragments (Chilton et al. 1977; Merlo et al. 1980). Second, the recombinant *E. coli* plasmids are present in multiple copies per cell whereas there is only one copy of the Ti-plasmid present per cell. Third, the use of recombinant molecules makes the preparation of large quantities of a specific DNA fragment feasible. Lastly, since only the Ti-plasmid portion of the recombinant molecule has homology to the T-DNA, whole plasmids can be radiolabeled to a very high specific activity for use as probes, and it is not necessary to extract bands from agarose gels.

ANALYSIS OF THE T-DNA IN VARIOUS OCTOPINE TUMOR GENOMES

The rationale used for analyzing the organization of T-DNA sequences in different tumor lines is essentially that described by Botchan, Topp, and Sambrook (1976) in showing the integration of SV40 DNA in rodent DNA. In brief, the total genomic DNA of a given tumor is digested with a restriction enzyme, and the large number of restriction fragments generated are separated by electrophoresis through agarose gels. The tumor DNA fragments are then transferred to nitrocellulose sheets by the Southern technique. The DNA bound to the nitrocellulose is allowed to hybridize with radiolabeled Ti-plasmid DNA, and the tumor DNA fragments containing Ti-plasmid sequences are located by autoradiography. Among the fragments will be those composed of only Ti-plasmid DNA (*internal fragments*), and they will have the same mobility (that is, molecular weight) as authentic Ti-plasmid restriction fragments generated by the digestion of the whole Ti-plasmid. In addition, there will be those fragments (*junction fragments*) composed of plasmid and plant sequences, assuming the T-DNA is integrated. These fragments will have homology to the Ti-plasmid but will have altered molecular weights when compared with authentic Ti-plasmid fragments. A comparison of the pattern of fragments with homology to the Ti-plasmid in the tumor digest with an authentic Ti-plasmid digest will show the extent of the plasmid sequences present in the tumor by the number of internal fragments. This analysis will provide a minimum estimate of the number of copies of the plasmid that are present by the number of junction fragments and their organization.

By using cloned plasmid fragments from the region of interest and allowing them to hybridize to genomic digests of four different tumor lines, we have been able to extend the observation of Merlo et al. (1980); we have

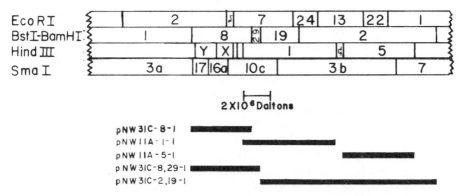

FIG. 8.1. Physical map of the T-DNA region of the Ti-plasmids of octopine-utilizing *A. tumefaciens* and a schematic representation of the cloned regions of the Ti-plasmid used in this study. Plasmids pTi-A6, pTi-B$_6$806, and pTi-15955 have been shown to be colinear through the T-region (Sciaky, Montoya, and Chilton 1978; Thomashow, Nutter, Montoya, Chilton, and Merlo, unpubl.). Employing the rationale used for construction of the physical map for pTi-B$_6$806 (Chilton, Montoya, Merlo, Drummond, Nutter, Gordon, and Nester 1978), we were able to establish the order of the restriction fragments for *Hind*III, *Bst*I or *Bam*HI (*Bam*HI is an isoschizomer of *Bst*I), and *Eco*RI digested Ti-plasmid. Small fragments have, in some cases, been given Greek letters since their precise number in the restriction digest has not been determined. The solid bars (**bottom** of the figure) represent regions of the Ti-plasmid that have been cloned into the *E. coli* plasmid pBR322 and used as probes. Details of the cloning and mapping of these fragments are presented elsewhere (Thomashow, Nutter, Postle, Chilton, Blattner, Powell, Gordon, and Nester 1980). (Reprinted from *Cell* 19 [1980]:729–39. © 1980 by MIT.)

begun to define the organization of the Ti-plasmid sequences in these tumor lines. A more detailed account of this work has been published elsewhere (Thomashow, Nutter, Montoya, Gordon, and Nester 1980; Thomashow, Nutter, Postle, Chilton, Blattner, Powell, Gordon, and Nester 1980).

EXTENT OF T-DNA

Merlo et al. (1980) showed that not all tumor lines studied contained the same T-DNA sequences. However, the technical limitations encountered prevented a precise description of the boundaries and the organization of the sequences. Genomic "blots" of three of the tumor DNA s studied in the previous work — A6S/2, B6806/E9, and 15955/1 — along with a fourth line — A277/5 (incited by pTi-B$_6$806) — were prepared as described by Thomashow, Nutter, Montoya, Gordon, and Nester (1980) and hybridized with labeled pTi-B$_6$806 plasmid or the two recombinant plasmids pNW31C-8,29 and pNW31C-2,19 (Fig. 8.1, **bottom**). Following autoradiography fragments showing homology to the entire Ti-plasmid were compared with the fragments showing homology to the two cloned Ti-fragment probes (Fig. 8.2). The number of fragments in a tumor line and the relative intensities of the bands were the same in both cases. This indicated that all the T-DNA is represented by the Ti-plasmid sequences present in the two clones. These data agree with the previously defined T-region (Chilton et al. 1977; Merlo et al. 1980).

The various tumor DNAs were digested with either *Hind*III, *Bst*I (an isoschizomer of *Bam*HI), *Eco*RI, or *Sma*I. DNA from these digests was treated as before, and the DNA bound to nitrocellulose sheets was hybridized with cloned fragments of the Ti-plasmid across the entire region of interest (see Fig. 8.1, **bottom**). Following autoradiography the pattern of internal plasmid fragments and junction fragments arising from the various restriction digests using probes from a specific region allowed us to draw several conclusions concerning the organization of the T-DNA in the four tumor lines (summarized in Fig. 8.3).

CORE T-DNA

All four lines studies contain a "core" region of plasmid-related sequences that extend from *Sma*I band 17 through *Eco*RI band 7 on the physical map. This DNA must contain the genes necessary for the maintenance of the transformed state since this is the only T-DNA present in A6S/2. However, as discussed below, at least one tumor line has an additional region of the Ti-plasmid present in its genome. Of particular interest is the fact that the core region present in all four octopine tumor lines contains the region of the Ti-plasmid termed the *common DNA* (Chilton, Drummond, Merlo, and Sciaky 1978). This supports the theory that the common DNA is directly involved in the transformation of the plant cell. In addition, it has been shown that complementary DNA (cDNA) made from total tumor poly(A)[+] RNA is homologous to the core T-DNA (Gelvin, unpubl.). Tumor polysomal messenger RNA (mRNA) can be hybridized to Ti-plasmid DNA from the region of the core T-DNA bound to nitrocellulose filters. When the bound mRNA is released by formamide treatment, it is able to direct the synthesis of at least one discrete polypeptide (McPherson, Nester, and Gordon, 1980). It is possible that this represents a regulatory protein that is involved in maintaining the tumorous state.

We have found that the left-hand boundary of the T-DNA in the four tumor lines studied was always within the 2.0 kb *Sma*I fragment 17. However, the precise point at which Ti-plasmid and plant sequences are joined in these lines can only be determined by sequencing the DNA through this region. In contrast, the right-hand boundary of the core region varies be-

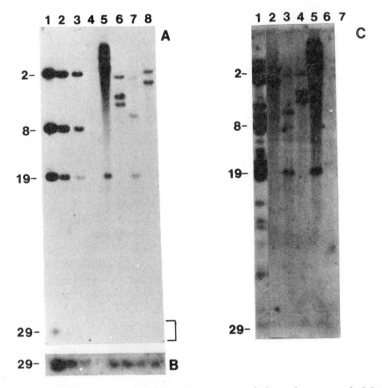

FIG. 8.2. Plant tumor restriction fragments with homology to radiolabeled Ti-plasmid or cloned portions of the Ti-plasmid DNA were digested with *Bst*I, subjected to electrophoresis through agarose gels, and transferred to nitrocellulose sheets by the Southern method (1975). Conditions for the preparation of the nitrocellulose filters, hybridization of radiolabeled probe to the DNA on the filters, washing of the filters, and autoradiography in the figures in this chapter are detailed elsewhere (Thomashow, Nutter, Postle, Chilton, Blattner, Powell, Gordon, and Nester 1980). Then 7 μg of plant or tumor DNA were digested and analyzed. Ti-plasmid reconstructions were prepared by the addition of Ti-plasmid that had been digested with the appropriate restriction enzyme to 7 μg of heterologous carrier DNA so that the plasmid sequences were present at the indicated "copy" per diploid tobacco genome (Chilton et al. 1977). (A) Plant tumor DNA restriction fragments having homology to radiolabeled cloned Ti fragments pNW31C-8,29 and pNW31C-2,19. Lanes 1, 2, and 3 are 5-, 2.5- and 1-copy reconstructions, respectively; 4, normal tobacco callus DNA; 5, B6806/E9 tumor DNA; 6, 15955/1 tumor DNA; 7, A6S/2 tumor DNA; 8, A277/5 tumor DNA. (B) A longer exposure of the area bracketed in A. (C) Plant tumor restriction fragments having homology to radiolabeled pTi-A6. Lane 1, 2.5 copy reconstruction; 2, A277/5 tumor DNA; 3, A6S tumor DNA; 4, 15955/1 tumor DNA; 5, B6806/E9 tumor DNA; 6, normal tobacco callus DNA; 7, a shorter exposure of 5. The positions of certain *Bst*I restriction fragments from the Ti-plasmid are indicated on the side. (Reprinted from *Cell* 19 [1980]:729–39. © 1980 by MIT.)

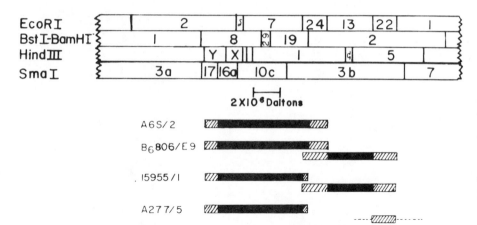

FIG. 8.3. Schematic representation of the T-DNA maintained in various octopine tumors. The **top** is the portion of the physical map of the octopine Ti-plasmids as described in Fig. 8.1. The **bottom** is a diagram of the T-DNA maintained in each tumor line studied. The solid lines indicate the regions of the plasmid that are present in their entirety as determined by the restriction analysis of the various tumor DNAs with the enzymes listed and the radiolabeled cloned Ti-plasmid fragments shown in Fig. 8.1. The hatched areas represent the regions where the precise boundaries of the T-DNA have not yet been determined. (Reprinted from *Cell* 19 [1980]:729–39. © 1980 by MIT.)

tween tumor lines. A6S/2 and B6806/E9 both have T-DNA that extends into *Bst*I fragment 2, since *Bst*I fragment 19 is present in its entirety (see Fig. 8.2, lanes 5 and 7; Fig. 8.3). On the other hand, 15955/1 and A277/5 both have the rightmost boundary of the core region terminating in *Bst*I fragment 19 (Fig. 8.2, lanes 6 and 8; Fig. 8.3). Again, the exact boundaries are not known at this time. We view as significant the finding that the ability of these four tumor lines to produce octopine (A6S/2 and B6806/E9 are producers; 15955/1 and A277/5 are not) correlates with the extent of the right-hand boundary of the core region of the T-DNA. Whether this is a region that codes for the opine synthesizing enzyme (Hack and Kemp 1980) or a regulatory protein for the synthesis of the enzyme is not known. Now that the opine synthesizing enzyme has been purified, antibodies can be produced that should be able to detect whether the enzyme is coded for by this region.

RIGHT-HAND T-DNA

In addition to the core DNA found in all tumors, we found that B6806/E9, 15955/1, and A277/5 contained DNA sequences from a region farther to the right in the physical map of the Ti-plasmid (see Fig. 8.3). The significance of these sequences is not known, though it is clear that they are not necessary for tumor maintenance since A6S/2 has none. A277/5 has only a small portion from this region, the boundaries of which have not been precisely determined. 15955/1 and B6806/E9 seem to have roughly the same extent of DNA from this region. However, B6806/E9 has these sequences in multiple copies of perhaps 20 to 30, but 15955/1 has only 1 to 2 copies of this region.

A more detailed analysis of the organization of this region in the A277/5 and 15955/1 tumors revealed a rather interesting feature. Whereas the right-hand region is not continuous to the core region on the physical map of the Ti-plasmid, we found that the data are consistent with the right-hand region being fused to the core region in the tumor genome. Presumably the DNA sequences lying between these regions have been deleted. An extensive rearrangement of the sequences seems unlikely in light of the co-linearity of the rest of the T-DNA with the Ti-plasmid. The multiple-copy region in the B6806/E9 tumor does not appear to be fused to the core region. However, the number of boundary fragments present in this region are so numerous that we cannot rule out the probability that there might be one copy of this region that is indeed fused to the core region. The number of junction fragments associated with this region roughly corresponds to the copy number of the central part of this region. In this regard it is most easily reasoned that the multiple-copy region was either inserted into a repetitive gene or amplified after integration.

OTHER FEATURES OF T-DNA IN OCTOPINE TUMORS

We find that there are a number of interesting observations that can be made with regard to the T-DNA found in the B6806/E9 and A277/5 tumors. Both tumors were incited by the same virulence plasmid (pTi-B$_6$806) in the same species of plant (*Nicotiana tabacum* var. Xanthi nc). Even though the left-hand boundary of the core DNA is in the same restriction fragment of the plasmid DNA, the molecular weight of the boundary fragment is different in the two tumors (see Fig. 8.4, lanes 3 and 6). Since we know the maximum variation that can be due to a different amount of *Sma* fragment 17 being present (no more than 2.0 kb), and it is clear that the boundary fragments vary by more than that amount, this variation must be due to the plant DNA portion of the fragment. Therefore, the Ti-plasmid must have been inserted into a different gene in these tumors. It is also clear that the two tumors do not have the same complement of T-DNA. Whether these differences are a result of different initial transformation events and/or subsequent rearrangements of the T-DNA is not yet known.

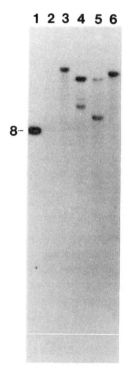

FIG. 8.4. Plant tumor DNA restriction fragments showing homology to the radiolabeled Ti-plasmid clone pNW31C-8. Plant tumor DNA, normal
plant DNA, or Ti-plasmid DNA was digested with
the restriction enzyme BstI. All conditions were the
same as outlined in Fig. 8.2. Lane 1, one-copy reconstruction; 2, normal tobacco callus DNA; 3, B6806/
E9 tumor DNA; 4, 15955/1 tumor DNA; 5, A6S/2
tumor DNA; 6, A277/5 tumor DNA. (Reprinted
from Cell 19 [1980]:729–39. © 1980 by MIT.)

With regard to A277/5, we have also compared the T-DNA of
A277/5 with the T-DNA present in the uncloned parent tumor line. We find
no evidence to suggest that within a given uncloned tumor line there are
cells with different complements of T-DNA. In addition, the data suggest
that the uncloned tumor DNA does not contain appreciable numbers of
normal cells that might be cross-fed by the tumorous cells.

INTEGRATION OF T-DNA

Autoradiographic analysis of the restriction fragments that demonstrated
homology to the Ti-plasmid revealed that all tumor lines had putative junction fragments with altered molecular weight compared with authentic Ti-
plasmid fragments (Fig. 8.4). It is felt that these fragments are the result of
the attachment of Ti-plasmid DNA to plant DNA since the results of other
experiments are not consistent with a large T-DNA replicon.

The above data argue strongly in favor of the interpretation that the
T-DNA is integrated into high molecular weight plant DNA. However,

more direct evidence was needed before it could be unequivocally stated that integration had occurred. We accomplished this by the cloning of restriction fragments comprising the entire genome of one of the tumor lines studied, A6S/2, into lambda phages. A detailed account of this work has been described elsewhere (Thomashow, Nutter, Postle, Chilton, Blattner, Powell, Gordon, and Nester 1980).

In brief, we subjected high molecular weight A6S/2 tumor DNA to partial digestion with the restriction enzyme *Eco*RI, selected the 15 to 20 kb fragments, and presented these fragments to *Eco*RI digested arms from lambda phage Charon 4A in the presence of T_4 DNA ligase (Blattner et al. 1977). The ligated mixture was packaged in vitro (Blattner et al. 1978), and the recombinant phages were screened (Benton and Davis 1977) for T-DNA sequences using the Ti-plasmid clones from the T-DNA region. After screening the recombinant phage library we detected three phages that contained DNA homologous to the T-region. DNA was isolated from these phages and restricted with *Eco*RI; the resulting fragments were separated by gel electrophoresis. The restriction fragments were visualized with UV light after staining with ethidium bromide (Fig. 8.5A). When the size of the fragments in the lambda clones are compared with the genomic fragments of A6S/2 tumor DNA that have homology to the Ti-plasmid (Fig. 8.5B), they appear to be the same. There are a total of four fragments visualized in the autoradiogram, two of which are internal fragments corresponding to Ti-plasmid fragments *Eco*RI 7 and δ. The remaining bands were putative junction fragments that arose from the integration of the Ti-plasmid DNA into the plant genome.

To show that the putative boundary fragments contained both plant and plasmid sequences — and thus were covalently linked to each other — the DNA from the recombinant phages was digested with *Eco*RI and subjected to electrophoresis. The resulting restriction fragments were transferred to nitrocellulose filters, and the DNA on the filters was allowed to hybridize to either radiolabeled Ti-plasmid clones or normal plant DNA. Fig. 8.6 shows that the putative boundary fragments hybridized to both Ti-plasmid DNA and normal plant DNA. This provided direct evidence that the T-DNA is attached to plant DNA.

We have subcloned the two ends of the A6S/2 tumor DNA (*left end* [LE] [pCG8.8] and *right end* [RE] [pCG3.0$\alpha\beta$]) into pBR325 so that they might be more carefully studied. Radiolabeled clones containing the LE or RE of A6S/2 tumor DNA were allowed to hybridize to nitrocellulose filters containing either Ti-plasmid digests or normal plant DNA digests on them. As expected, both end-fragment clones had sequences homologous to the Ti-plasmid. Of interest were the results obtained when the end-fragment clones were hybridized to digests of normal plant DNA as shown in Fig. 8.7. The LE clone has homology to primarily one band in the normal plant

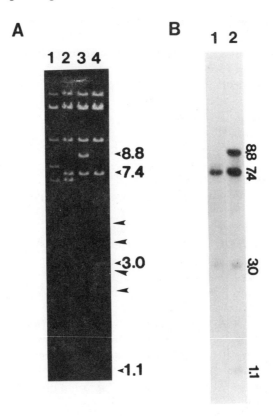

FIG. 8.5. Comparison of restriction fragments from Charon 4A clones of A6S/2 tumor DNA to restriction fragments of A6S/2 DNA with homology to radiolabeled Ti-plasmid clones. A6S/2 tumor DNA was digested with *Eco*RI and the 15 to 20 kb fragments were cloned into Charon 4A, packaged in vitro, and screened for inserts having homology to the cloned Ti-plasmid sequences from the T-region as described (Thomashow, Nutter, Postle, Chilton, Blattner, Powell, Gordon, and Nester 1980). (A) DNAs from the three Charon 4A recombinant phages were digested with *Eco*RI, and the resulting fragments were separated by electrophoresis through an agarose gel. The fragments were visualized after staining the gel in 1 μg/ml ethidium bromide and illumination on UV light box. Lane 1, Charon 4A DNA; 2, pCGλ34 DNA; 3, pCGλ31 DNA; 4, pCGλ5 DNA. The fragments attributed to A6S/2 tumor DNA are indicated by arrows on the side of the gel, and the sizes in kb of the major fragments are noted. (B) Restriction fragments resulting from the digestion of 7 μg A6S/2 tumor DNA with *Eco*RI demonstrating homology to radiolabeled Ti-plasmid clones (see Fig. 8.1). All conditions are as described in Fig. 8.2. Lane 1, fragments showing homology to radiolabeled pNW31C-2,19; 2, fragments showing homology to radiolabeled pNW31C-8,29. The sizes of the fragments in kb are indicated on the side of the autoradiogram.

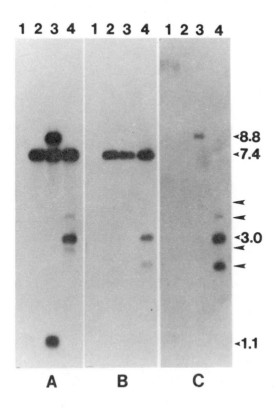

FIG. 8.6. Restriction fragments generated from A6S/2 tumor DNA cloned in lambda phage Charon 4A showing homology to radiolabeled clones of the Ti-plasmid or normal plant DNA. *Eco*RI was used to digest 0.5 μg of DNA isolated from Charon 4A or the three clones identified in Fig. 8.5. The fragments were separated by electrophoresis through agarose gels and transferred to nitrocellulose sheets. All conditions are as previously described. (A) Restriction fragments from DNA isolated from Charon 4A phage showing homology to radiolabeled pNW31C-8,29. Lane 1, Charon 4A DNA; 2, pCGλ34 DNA; 3, pCGλ31 DNA; 4, pCGλ5 DNA. (B) The same as in A except the radiolabeled probe was pNW31C-2,19. (C) The same as in A except the radiolabeled probe was total DNA isolated from tobacco callus. In all three figures, the minor fragments indicated only by arrows are thought to be due to recombinational events in the phage. This phenomenon has been reported to occur also in murine sarcoma virus (MSV) virus DNA cloned in lambda (Tronick et al. 1980). The sizes of the major fragments showing homology to the Ti-plasmid in A6S/2 tumor DNA are indicated in kb along the side of the autoradiogram in C.

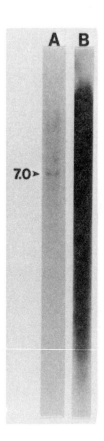

FIG. 8.7. Restriction fragments from normal
plant DNA demonstrating homology to radiolabeled
clones containing the LE or RE from A6S/2 tumor DNA.
*Eco*RI was used to digest 1 μg of normal tobacco callus
DNA. The fragments were separated by electrophoresis
through agarose gels and transferred to nitrocellulose
sheets. The DNA on the sheets was allowed to hybridize
to radiolabeled clones prepared from the A6S/2 tumor
DNA cloned in Charon 4A as described (Thomashow,
Nutter, Postle, Chilton, Blattner, Powell, Gordon, and
Nester 1980). All conditions are as described previously.
(**A**) Normal plant DNA restriction fragments demonstrating homology to the cloned LE (pCG8.8) of A6S/2
tumor DNA. The size in kb of the distinct fragment
demonstrating homology is indicated on the side of the
autoradiogram. (**B**) Normal plant DNA restriction fragments demonstrating homology to the cloned RE
(pCG3.0αβ) of A6S/2 tumor DNA.

digest with slight homology to a broad range of restriction fragments in the
plant. The RE clone has very strong homology to plant restriction fragments throughout the entire digest, suggesting that the cloned plant DNA
contains a highly repetitive family of genes. This is substantiated by comparing the intensity of the LE fragment (lane 3, 8.8 kb) to the RE fragment
(lane 4, 3.0 kb) in Fig. 8.6C. The present data do not allow us to say whether the T-DNA is integrated into a repetitive family of plant DNA or into a
unique region that is close to a repetitive region. It seems that the repetitive
DNA sequences that are associated with the ends of the T-DNA have no
homology to each other since radiolabeled LE does not hybridize to the RE
under the conditions of our hybridization.

SUBCELLULAR LOCATION OF T-DNA

The T-DNA of the A6S/2 tumor line is integrated into nuclear DNA. This is
suggested by the observation that both end fragments in the A6S/2 tumor

DNA have homology to a broad range of restriction fragments from the plant DNA. Only nuclear DNA has a complexity great enough to give such a broad range of restriction fragments. Mitochondrial and chloroplast DNA are much less complex than nuclear DNA and give discrete bands upon digestion with restriction enzymes. More important, we find that there is no homology between the LE or RE and chloroplast or mitochondrial DNA. Therefore, the plant sequences must be of nuclear origin.

CONCLUSION

There are a number of important observations that can be made relative to the molecular biology of the Ti-plasmid and/or to the possible use of the Ti-plasmid as a cloning vehicle for the insertion of foreign DNA into plants.

Molecular Biology of T-DNA

First, the entire T-region is not necessary for the maintenance of the transformed state. It is evident that the T-DNA found in A6S/2 is adequate for tumor maintenance — therefore, the genes controlling the transformed state must be found in this region. In addition, this region contains the sequences that have been called common DNA, and this is consistent with this region's playing a direct role in virulence (Chilton, Drummond, Merlo, and Sciaky 1978; DePicker, Montagu, and Schell 1978). However, the minimum amount of T-DNA necessary for maintenance of the transformed state is not known. Koekman et al. (1979) have shown that deletions made in an octopine virulence plasmid that extend into the T-region from the right to a point in *Bst* fragment 19 still lead to the formation of tumors when bacteria containing these plasmids are inoculated into a plant. Members of our group are currently investigating the T-DNA in tumors that result from the transformation of plant protoplasts by whole agrobacteria or naked Ti-plasmid. The extent and organization of T-DNA in these tissues could lead to a more precise definition of the T-DNA sequences necessary for transformation.

Second, the extent and organization of T-DNA in tumors is not the same even when bacteria containing the same virulence plasmid are inoculated into the same species of plant. In fact, it does not seem that the T-DNA is inserted into the same plant gene. The significance of this observation is not known at this time. Botchan, Topp, and Sambrook (1976) found that SV40 DNA integrated at random points in the rat genome.

And third, there appears to be at least one preferred region of the Ti-plasmid for insertion, since the left-hand end of the core region in all four tumors is within the 2.0 kb *Sma* fragment 17 of the Ti-plasmid. The right-hand end of the core region, though, does not share this trait. From the information that we have at this time it does not seem that the Ti-plasmid acts like the transposable elements currently known to exist in prokaryotes (Kleckner 1977). It seems that there are no long repeats, either direct or inverted, at the LE or RE of the A6S/2 T-DNA since the cloned ends show no homology to each other as do most of the transposable elements. However, we cannot rule out the presence of small repeats at these points that are beyond our levels of detection or the possibility that the right-hand repeat might have been lost during deletion or rearrangement of the T-DNA.

Possibilities for Future Use of the Ti-Plasmid as a Cloning Vehicle

The T-DNA is integrated into the plant nuclear DNA. If genetic engineering of plants is to become a reality, the foreign genes must be stably maintained in the plant genome. Foreign genes that are integrated into genomic DNA would be more likely to be maintained in the cell from generation to generation.

It is also possible that when the minimum amount of T-DNA necessary for maintenance of the transformed state is defined, selected genes could be attached to the transforming region and would be integrated and expressed. Selection of engineered plant cells could be accomplished simply by determining those cells that were transformed.

Finally, as the manner in which one might genetically engineer plant cells becomes better understood, the transformed phenotype will become less desirable. This is particularly true if the plants to be altered are ornamental or food crops. It will then be necessary for molecular biologists to be able to define what the properties of the Ti-plasmid are that allow it to integrate into the host genome. The precise steps that lead to the integration of the Ti-plasmid are not known at this time. It has been shown that mutations in the T-region or other regions of the Ti-plasmid can alter or abolish virulence (Garfinkel and Nester 1980), but the functions affected have not been determined. If we can define the functions that are essential for integration, then it is conceivable that the Ti-plasmid can be engineered so that the transforming genes are deleted from the portions necessary for integration. Desirable genes could then be inserted in place of the transforming genes. Of course, a more elaborate means of selecting the engineered plants will have to be developed since the plants will have a normal phenotype.

Although more work remains to be done, we feel that the prospects for utilizing the Ti-plasmid for the introduction of foreign genes into plants are possible and deserve continued attention.

REFERENCES

Benton, W. E., and Davis R. W. 1977. Screening λgt recombinant clones by hybridization to single plaques in situ. *Science* 196:180–82.

Blattner, F. R.; Blechl, A. E.; Denniston-Thompson, K.; Faber, H. E.; Richards, J. E.; Slightom, J. L.; Tucker, P. W.; and Smithies, O. 1978. Cloning human fetal γ globin and mouse α-type globin DNA: preparation and screening of shotgun collections. *Science* 202:1279–84.

Blattner, F. R.; Williams, B. G.; Blechl, A. E.; Thompson, K. D.; Faber, H. E.; Furlong, L. A.; Grundwald, D. J.; Keifer, D. O.; Moore, D. O.; Schumm, J. W.; and Smithies, O. 1977. Charon phages: safer derivatives of bacteriophage lambda for DNA cloning. *Science* 196:161–69.

Bomhoff, G.; Klapwijk, P. M.; Kester, H. C. M.; Schilperoort, R. A.; Hernalsteens, J. P.; and Schell, J. 1976. Octopine and nopaline synthesis and breakdown genetically controlled by a plasmid of *Agrobacterium tumefaciens. Mol. Gen. Genet.* 145:177–81.

Botchan, M.; Topp, W.; and Sambrook, J. 1976. The arrangement of Simian virus 40 sequences in the DNA of transformed cells. *Cell* 9:269–87.

Braun, A. C. 1947. Thermal studies on the factors responsible for tumor initiation in crown gall. *Am. J. Bot.* 34:234–40.

Chilton, M. D.; Drummond, M. H.; Merlo, D. J.; and Sciaky, D. 1978. Highly conserved DNA of the Ti-plasmids overlaps T-DNA maintained in plant tumors. *Nature* (London) 275:147–49.

Chilton, M. D.; Drummond, M. H.; Merlo, D. J.; Sciaky, D.; Montoya, A. L., Gordon, M. P.; and Nester, E. W. 1977. Stable incorporation of plasmid DNA into higher plant cells: the molecular basis of crown gall tumorigenesis. *Cell* 11:263–71.

Chilton, M. D.; Montoya, A. L.; Merlo, D. J.; Drummond, M. H.; Nutter, R.; Gordon, M. P.; and Nester, E. W. 1978. Restriction endonuclease mapping of a plasmid that confers oncogenicity upon *Agrobacterium tumefaciens* strain B6-806. *Plasmid* 1:254–69.

DePicker, A.; Van Montagu, M.; and Schell, J. 1978. Homologous DNA sequences in different Ti-plasmids are essential for oncogenicity. *Nature* (London) 275: 150–53.

Drummond, M. H.; Gordon, M. P.; Nester, E. W.; and Chilton, M. D. 1977. Foreign DNA of bacterial plasmid origin is transcribed in crown gall tumours. *Nature* (London) 269:535–36.

Garfinkel, D. J., and Nester, E. W. 1980. *Agrobacterium tumefaciens* mutants affected in crown gall tumorigenesis and octopine catabolism. *J. Bacteriol.* 144: 732–43.

Gordon, M. P. Forthcoming. Tumor formation in plants. In *The biochemistry of plants: a comprehensive treatise*, ed. A. Marcus. New York: Academic Press.

Gurley, W. B.; Kemp, J. D.; Albert, M. J.; Sutton, D. W.; and Callis, J. 1979. Transcription of Ti-plasmid-derived sequences in three octopine-type crown gall tumor lines. *Proc. Nat. Acad. Sci. (USA)* 76:2828–32.

Hack E., and Kemp, J. D. 1980. Purification and characterization of the crown gall specific enzyme octopine synthase. *Plant Physiol.* 65:949–55.

Kleckner, N. 1977. Translocated elements in prokaryotes. *Cell* 11:11–23.

Koekman, B. P.; Ooms, G.; Klapwijk, P. M.; and Schilperoort, R. A. 1979. Genetic map of an octopine Ti-plasmid. *Plasmid* 2:347–57.

Lippincott, J. A., and Lippincott, B. B. 1975. The genus *Agrobacterium* and plant tumorigenesis. *Ann. Rev. Microbiol.* 29:377–405.

McPherson, J. C.; Nester, E. W.; and Gordon, M. P. 1980. Proteins encoded by *Agrobacterium tumefaciens* T-DNA in crown gall tumors. *Proc. Nat. Acad. Sci. (USA).* 77:2666–70.

Merlo, D. J.; Nutter, R. C.; Montoya, A. L.; Garfinkel, D. J.; Drummond, M. H.; Chilton, M. -D.; Gordon, M. P.; and Nester, E. W. 1980. The boundaries and copy numbers of Ti-plasmid T-DNA vary in crown gall tumors. *Mol. Gen. Genet.* 177:637–44.

Sciaky, D.; Montoya, A. L.; and Chilton, M. -D. 1978. Fingerprints of *Agrobacterium tumefaciens* plasmids. *Plasmid* 1:238–53.

Smith, E. F., and Townsend, C. O. 1907. A plant tumor of bacterial origin. *Science* 25:671–73.

Southern, E. M. 1975. Detection of specific sequences among DNA fragments separated by gel electrophoresis. *J. Mol. Biol.* 98:503–17.

Thomashow, M. F.; Nutter, R.; Montoya, A. L.; Gordon, M. P.; and Nester, E. W. 1980. Integration and organization of Ti-plasmid sequences in crown gall tumors. *Cell* 19:729–39.

Thomashow, M. F.; Nutter, R.; Postle, K.; Chilton, M. -D.; Blattner, F. R.; Powell, A.; Gordon, M. P.; and Nester, E. W. 1980. Recombination between higher plant DNA and the Ti-plasmid of *Agrobacterium tumefaciens*. *Proc. Nat. Acad. Sci. (USA)* 77:6448–52.

Tronick, S. R.; Robbins, K. C.; Canasni, E.; Devare, S. G.; Anderson, P. R.; and Aaronson, S. A. 1980. Molecular cloning of Moloney Murine sarcoma virus: arrangement of virus-related sequences within the normal mouse genome. *Proc. Nat. Acad. Sci. (USA)* 76:6314–18.

Van Larebeke, N.; Engler, G.; Holsters, M.; Van Den Elsacker, S.; Zaenen, I.; Schilperoort, R. A.; and Schell, J. 1974. Large plasmid in *Agrobacterium tumefaciens* essential for crown gall-inducing ability. *Nature* 252:169–70.

Watson, B.; Currier, T. C.; Gordon, M. P.; Chilton, M.-D.; and Nester, E. W. 1975. Plasmid required for virulence of *Agrobacterium tumefaciens*. *J. Bacteriol.* 123:255–64.

Zaenen, I.; Van Larebeke, N.; Teuchy, H.; Van Montagu, M.; and Schell, J. 1974. Supercoiled circular DNA in crown gall-inducing *Agrobacterium* strains. *J. Mol. Biol.* 86:109–27.

9

TRANSPOSON-INDUCED SYMBIOTIC MUTANTS OF *RHIZOBIUM MELILOTI*

Sharon R. Long
Harry M. Meade
Susan E. Brown
Frederick M. Ausubel

INTRODUCTION

Bacteria of the genus *Rhizobium* establish symbioses with plants of the family Leguminosae; within this general pattern, more restricted specificities exist between particular species of *Rhizobium* and particular genera or species of legume. The structure formed by the two interacting partners is a root nodule, within which some plant-derived cells are packed with *bacteroids*, a differentiated form of the infecting bacteria. The mature symbiotic nodule fixes atmospheric nitrogen into ammonia, supplying the plant with metabolically usable nitrogen. In our laboratory we have been studying the interaction of *Rhizobium meliloti* and one of its host plants, alfalfa (*Medicago sativa* L.)

Much is still unknown about the symbiosis—for example, it is not known whether the general restriction of rhizobial symbioses to legumes as opposed to other plants and the finer species specificity are two expressions of one selective mechanism or derive from two or more different ones. The signal that causes the cells of the root cortex to proliferate, forming the nodule, has not been identified; and it is not at present known whether any

We would like to acknowledge Gary Ruvkun for the construction of pRmR2 and pRmR3 and to thank him and Rod Riedel for their substantial contributions to improving the techniques used here for DNA analysis and cloning. Harry M. Meade and Sharon R. Long were supported by postdoctoral fellowships from the National Science Foundation (NSF) and the National Institutes of Health. Additional support for this project was provided by NSF grant PCM78-06834 to Frederick M. Ausubel.

genetic information passes between the invading *Rhizobium* and the host cells.

In Table 9.1 we briefly summarize the major stages in the development of a symbiotic root nodule (see also Vincent [1974,1980] for a more detailed description of the symbiotic sequence.) Presumably, some surface interaction of the two partners is involved in their initial recognition; the earliest visible sign of infection is the presence of curled root hairs and the growth of an infection thread through the infected root hair into the main body of the root. As the infection thread penetrates, it forms branches and the contained bacteria proliferate. At about this stage the host cells in the inner root cortex begin to divide. Eventually, various branches of the infection thread invade some host cells and release bacteria. The bacteria undergo morphological and biochemical differentiation; in their mature symbiotic state they are referred to as bacteroids. The host cell, too, shows changes in protein composition (Legocki and Verma 1980) and ultrastructure (Libbenga and Bogers 1974; Mosse, 1964).

The development of the symbiotic nodule can be viewed as a series of steps, each one involving the interaction of host plant and bacterium. In its most simple form, we analogize this to a metabolic pathway in which each genetically controlled step produces an intermediate structure, which, in turn, is used as substrate for a further conversion. The final product – in this case, a nodule – is formed through the functioning of the entire pathway, and a mutation in any one step may be sufficient to prevent the formation of the final product. And, as in a metabolic pathway, a lesion in any one step may result in the accumulation of the intermediate that is produced just prior to that step.

TABLE 9.1
Steps in Symbiosis – Developmental Pathway

---

1. Recognition of bacteria and plant.

2. Invasion of root hair by bacterium; formation of infection thread.

3. (a) Penetration and branching of infection thread; proliferation of bacteria within infection thread.
   (b) Host cells (root inner cortex) begin to divide.

4. Release of bacteria from infection thread into some cells of growing nodule. Differentiation of bacteria and plant cells:
   (a) Bacteria differentiate morphologically into bacteroids. New proteins, including nitrogenase.
   (b) Host-cell ultrastructural changes (membranes, cell organelle arrangement). New proteins, for example, leghemoglobin.

Source: Compiled by the authors. See also Vincent (1980) for more details.

If different bacterial genes affect different steps in the symbiotic developmental pathway, then this leads to predictions of the type of symbiotic phenotype that would result from bacterial mutations, such as the following. A mutation affecting a function(s) required in an early step(s) of the developmental pathway—recognition or binding, for example—would have the result that no nodule would form at all. A bacterial strain carrying a mutation affecting only the enzyme nitrogenase might produce nodules on its host plant, but the nodules would not fix nitrogen. Thus, an array of symbiotic phenotypes can be expected—from no nodule through abnormal or incomplete nodules to fully formed but ineffective symbiotic nodules—according to the step in the nodule's development in which the bacteria failed to interact properly with the host. It is also likely that host genes exist that, when mutant, cause failure of specific steps in the pathway to nodule formation (Bergersen and Nutman 1957; Vincent 1974, 1980).

We have isolated mutants of R. meliloti by transposon mutagenesis and among them have found strains that produce no nodules (or abnormal nodules) and that form nodules incapable of nitrogen fixation. We discuss the generation of these symbiotic mutations, the physical and genetic methods used to analyze them, and the effect of these mutations on the development of the symbiosis.

SYMBIOTIC MUTANTS OF *RHIZOBIUM*

Preliminary genetic studies of symbiotic nitrogen fixation have been carried out in several species of *Rhizobium*. Maier and Brill (1976) obtained five nodulation-defective mutants of R. japonicum after screening 2,500 mutagenized clones for nodulating ability. In a study with R. leguminosarum, Beringer, Johnston, and Wells (1977) found that after mutagenesis with N-methyl-N'-nitro-N-nitrosoguanidine (NTG) 3 percent of survivors were nodulation defective and nonauxotrophic while another 3 percent were auxotrophs with nodulation deficiencies. Other laboratories have studied the effect on symbiotic phenotype of bacterial mutations conferring auxotrophy (Dénarié, Truchet, and Bergerson 1976; Pankhurst and Schwinghamer 1974; Schwinghamer 1977) and drug or antimetabolite resistance (Amarger 1975; Hendry and Jordan 1969; Schwinghamer 1968). An important finding of many of these studies was that when such mutations affect the symbiosis, they often cause arrest at a particular stage. For example, an ade⁻thi⁻auxotroph of R. leguminosarum (Pankhurst and Schwinghamer 1974) forms a normal number of infected root hairs, and small nodules are initiated by the plant, but no bacteria are released from the infection threads. A leu⁻ mutant of R. meliloti is also blocked at the bacterial release stage (Truchet, Michel, and Dénarié 1980).

A problem in beginning genetic studies in such a system as this is that it is not clear in advance which kind of bacterial mutation will affect the symbiosis. In the absence of such knowledge it is not possible to select for phenotypes in the free-living bacterium that will affect all the various stages of symbiotic development. Therefore, it is necessary to generate mutants at random in *Rhizobium* and to screen each of these potentially mutant strains for an effect on the symbiotic process. One technical constraint in adopting a screening strategy is that thousands of bacterial strains must be individually tested on plants, so that the space taken up by the plants may become a problem. We have used alfalfa at the seedling stage for our screening, because each plant can be grown in a test tube (18 × 150 mm) and tubes can be assembled together in boxes of 200 for space-efficient growth.

A second constraint that would accompany the use of classical bacterial genetics in such a project comes from the possible lack of expression in free-living cells of the genetic loci causing mutant symbiotic phenotypes. Revertants would also not be detectable in culture, nor would it be possible to map mutants without going through a plant test for all the progeny of each bacterial cross. To overcome this problem we have used the technique of transposon mutagenesis. This has been used in many bacteria (see reviews by Kleckner [1977] and Berg [1977]) and was first applied to a *Rhizobium* species by John Beringer, Andrew Johnston, and their colleagues studying *R. leguminosarum* (Beringer et al. 1978). The transposon Tn5, which encodes resistance to kanamycin and neomycin, is capable of transposing from one location in a cell's DNA to another. When it inserts into a new location, it interrupts the function of whatever gene might be there and confers upon the region a new resistance to kanamycin/neomycin. We adopted the method of Beringer et al. (1978) to transfer Tn5 into *R. meliloti*. This method, called suicide plasmid mutagenesis, involves conjugating into the target cell a plasmid that is incapable of surviving in its new host cell and that carries a copy of the transposon (Van Vliet et al. 1978). By selecting for the resistance encoded by the transposon, colonies can be selected in which the transposon has survived the "suicide" of its original plasmid by hopping into the host cell's own deoxyribonucleic acid (DNA) and replicating along with it. When the transposon inserts into a host gene, it causes a null mutation for that particular gene; such mutants will be found as antibiotic-resistant exconjugants of a transposon suicide mutagenesis. Each mutation, whatever its other phenotype, is marked in the free-living bacteria by the antibiotic resistance, and its position can be mapped by conjugation and linkage analysis, as any other resistance locus would be.

We used the suicide plasmid pJB4JI, which contains Tn5, to mutagenize *R. meliloti* strain Rm1021, a streptomycin-resistant derivative of SU47. All Tn5-containing exconjugants were identified and maintained on plates containing both streptomycin (500 μg/ml) and neomycin (50 μg/ml).

All mutant colonies were purified to single colonies and tested for auxotrophy and for residual suicide plasmids before being individually inoculated onto alfalfa seedlings to examine their symbiotic phenotype.

Over 6,000 individual *R. meliloti* mutants were included in our initial project to screen for mutations in symbiotic and nitrogen-fixation genes. Among the 49 symbiotic mutants found were 29 in which internal rearrangements of *R. meliloti* DNA, or persistence of parts of the suicide plasmid, made genetic interpretation complicated (Long et al., submitted; Meade et al., submitted). We will not discuss those cases here, but we will concentrate instead on several of the other mutants in which a single transposition of Tn5 into the *R. meliloti* genome has caused a change in the bacterium's ability to form a successful symbiotic nodule. In the following section we describe how we are characterizing these mutants from a developmental standpoint and briefly discuss the physical and genetic analysis of the mutant strains.

PHENOTYPES OF THE MUTANT STRAINS

As described above, symbiotic mutations will have various phenotypes, according to which step in nodule development is affected by the genetic lesion in the bacterium. One type of mutant is "unreactive Nod⁻," whose phenotype in the initial screen was failure to produce any nodules (Nod⁻). Upon microscopic examination we discovered that this type of mutant seems not to react with root hairs: in the conditions suggested by Fåhraeus (1957), with modifications by Nutman (1971), no curled root hairs are seen, as would be in a normal *Rhizobium* inoculation. Using the pathway analogy, we would say that the mutation is such a strain affected a step prior to root hair curling, which therefore must be a very early step in the host-bacterial interaction. One possibility is that the bacterium has either lost one of its needed binding components or lost the ability to stimulate the host to react properly.

A second class of mutant had the same overall phenotype in the initial screen, that is, Nod⁻. However, when inoculated onto plants in a Fåhraeus preparation, these mutants could be observed to react with root hairs: "shepherd's crooks" (the normal type of curled root hair) were found, along with striking, often bizarre-looking, branched and misshapen root hairs. We have called this class reactive Nod⁻ mutants. In these it would appear that a step between root hair curling and the stimulation of the host to produce a nodule has been affected by the mutation.

Other *R. meliloti* mutations results in the formation of nodules that are ineffective (Nod⁺ Fix⁻). The nodules are not always completely normal, they often grow much longer than those formed by wild type *R. meliloti* and are found in unusually large numbers, sometimes in clusters.

One method of analysis being performed on these mutants is examination of the ineffective nodules by scanning electron microscopy (SEM) (Hirsch and Long, unpubl.). Studies carried out so far indicate that several phenotypes may be found among the mutant nodules; in some no mature, elongated bacteroids can be seen. In these mutants the genetic defect seems to affect a step prior to the morphological differentiation of the bacteria. On the other hand, in some other Nod⁺ Fix⁻ mutants, the bacteroids appear to be morphologically normal: the lesion here must occur in some very late function, perhaps affecting the final biochemical differentiation or bacteroid and/or host cell. One mutant that forms normal-looking bacteroids apparently shows premature senescence of bacteroid and host-cell structures.

GENETIC STUDIES

There are two main goals of our genetic studies on symbiotic mutations in *R. meliloti*: first, to map the location on the genome of the Tn5 insertion sites by linkage analysis and, second, to establish that the Tn5 insertion is the cause of the symbiotic mutation by showing 100 percent linkage between the Tn5-coded drug resistance and the symbiotic lesion.

Mapping the location of Tn5 inserts is accomplished by conjugating the mutant containing Tn5, which is wild type in its other loci, into a recipient that carries multiple auxotrophic mutations and drug resistances. Conjugation does not seem to occur naturally in *R. meliloti* but will be promoted by the presence in the donor bacterium of the P-type plasmid RP4 or one of its close relatives (Kondorosi et al. 1977; Meade and Singer 1977). Under these conditions a segment of donor DNA is carried across into the recipient where it replaces the recipient's homologous segment. A rough estimate of the position of a particular Tn5 insert can be obtained by recording the relative number of times it is carried into a recipient along with each of the metabolic and resistance loci. Fig. 9.1 shows the approximate map locations of the loci mapped so far, relative to four auxotrophic loci (*pan*, *trp*, *his*, and *pyr*), and two chromosomally linked resistances (novobiocin and rifampicin). It is apparent that the various genes are not tightly clustered since they map at widely distant locations.

Some of the mutants appear to map in the same region—for example, 1088 and 1029 near the *pan* locus. However, with the conjugation mapping technique we can establish only approximate map locations. Therefore, the map shown may change as finer analyses are carried out, perhaps with physical mapping techniques.

FIG. 9.1. Linkage map of the *R. meliloti* chromosome showing approximate positions of Tn5-induced symbiotic mutations. Map arrangement follows the convention suggested by Kondorosi et al. (1980). Map positions were determined by measuring cotransfer frequency of neomycin resistance (coded for by Tn5) with metabolic or auxotrophic loci during R-factor mediated conjugation (Meade and Signer 1977). The R-factor used in these studies was pGM102, a kanamycin-sensitive derivative of RP4 (Meade et al., submitted). Note: 1027 is a Nod⁻ mutant; all others have a Fix⁻ phenotype.

    One useful application of the conjugation technique, however rough, is the second goal mentioned above: to establish the linkage between Tn5 and the symbiotic phenotype. This is tested by crossing a Tn5-carrying symbiotic mutant into a marked recipient strain that is symbiotically normal. All progeny in which the Tn5-carrying segment of donor DNA has been transferred into the recipient cell are saved and individually inoculated onto plants. If the Tn5 insertion was the cause of the donor strain's symbiotic deficiency, then all the progeny that received that segment of donor DNA will become symbiotically defective.

    This experiment has been carried out with three Fix⁻ mutants and one Nod⁻ mutant; in each case, the progeny receiving the donor's Tn5 were found to be symbiotically deficient. This strongly implies that the symbiotic mutation in these strains is due to Tn5. However, since in most conjugations large segments of donor DNA are transferred to the recipient cell, the mating data alone cannot be regarded as firm proof that Tn5 insertion is the unique cause of the symbiotic mutations in our strains. Experiments are currently under way that involve cloning from each mutant strain the individual restriction fragment of DNA that contains the Tn5 insert and its immediate surrounding DNA. Using recombinant conjugative plasmids, these individual fragments will be transferred and recombined into otherwise wild type *R. meliloti* cells, which will allow a much more stringent test of the link between Tn5 insertion and symbiotic mutation in these strains.

PHYSICAL ANALYSIS

Because Tn5 has no site for the restriction enzyme *Eco*RI, its insertion into a particular *Rhizobium Eco*RI fragment creates a new fragment larger by 5,700 base pairs (Fig. 9.2). Two straightforward kinds of analysis can be performed on the DNA of mutant *R. meliloti* strains to physically charac-

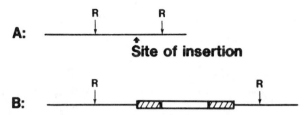

FIG. 9.2. Insertion of transposable resistance element Tn5 into a section of DNA from *R. meliloti*. Tn5 is approximately 5,700 base pairs long and consists of a central region where the kanamycin/neomycin resistance gene is located, flanked by two arms whose sequences are inverted repeats (shaded area). The Tn5 element has no restriction site for the enzyme *Eco*RI. When Tn5 inserts into an *R. meliloti* genomic site, here delimited as an *Eco*RI fragment of 4.0 kb, the size of the *Eco*RI fragment containing that site is increased by 5.7 kb.

terize the mutation. Both involve cutting total *R. meliloti* DNA with *Eco*RI and electrophoresing the fragments on an agarose gel to separate them. The separated fragments are then transferred to a nitrocellulose filter using the method of Southern (1975).

In each analysis such a filter is hybridized with a particular radioactively labeled probe. First, the filter can be probed with purified [^{32}P]-Tn5 DNA. In the DNA from each mutant the Tn5 DNA will hybridize with any fragment where Tn5 has inserted. Thus, it can be determined whether a mutant has only one, or more than one, Tn5 insert, and the size(s) can be estimated by comparison with marker fragments of known size. In Fig. 9.3 the autoradiogram from such a hybridization is shown. It can be seen that only one fragment in each sample hybridized to the labeled Tn5, implying a single insert in each mutant.

A second analysis can be carried out to test whether any particular fragment has been "hit" by Tn5, as follows. The DNA of a particular region from wild type *R. meliloti* is cloned. Then it is radioactively labeled and hybridized with filters onto which *Eco*RI-digested DNA of mutant strains has been transferred. If in any one mutant Tn5 happened to insert in the fragment represented by that cloned sequence, then the wild type sequence will hybridize to a larger fragment in that mutant and the band on the autoradiogram will be in a higher position.

We did this experiment using two fragments from wild type *R. meliloti*, which were cloned by Gary Ruvkun in this laboratory (Ruvkun and Ausubel 1981). One sequence was cloned by homology with the cloned nitrogenase structural genes *D* and *H* from *Klebsiella pneumoniae* (Cannon, Riedel, and Ausubel 1979), and the other sequence was immediately adjacent to it; these two cloned fragments, pRmR2 and pRmR3, are represented in Fig. 9.4. It can be seen that pRmR2 has a length of 3.9 kb and pRmR3 a length of 5.0 kb. When they are mixed together, labeled and probed onto wild type DNA, there are two bands of hybridization at 3.9 and 5.0 kb (Fig. 9.5, lane a). As described above, if a mutant strain has a Tn5 insertion in either of the two sequences used as probes, the hybridization to one of the two bands will be seen in a new location; such a case is found in mutant strain RmH1128 (Fig. 9.5, lane b). In this mutant, obtained by the suicide plasmid mutagenesis and screening, the sequence represented in plasmid pRmR3 has received a Tn5 insert, making the fragment size larger.

We can see, therefore, that a region that we might expect to be associated with nitrogen fixation because it is adjacent to the *Rhizobium* region homologous to *K. pneumoniae nif* DNA is one of the symbiotic genes "found" by a random Tn5 mutagenesis and screen for symbiotically defective phenotypes in *R. meliloti*. This is an important confirmation of the value of this approach and an encouragement for the further study of the other loci found in this experiment and the genetic regions adjacent to them.

CLONING OF DNA FROM MUTANT SYMBIOTIC GENES

The use of transposons to create mutations at symbiotic loci facilitates the cloning of DNA from those loci. Transposon 5 has no *EcoRI* site; therefore, in an *EcoRI* digest of total DNA from a Tn5-containing mutant, there will be a single fragment that contains Tn5 and two flanking *R. meliloti* sequences representing an *EcoRI* fragment from a symbiotic gene (refer to Fig. 9.2B). This composite *EcoRI* fragment can be cloned by ligation of total *EcoRI* fragments from a particular mutant into a vector plasmid such as pBR322 and transformation of those ligated recombinant molecules into an *Escherichia coli* cell placed under selection for kanamycin resistance. Only the recombinant plasmid that includes the *R. meliloti*::Tn5 fragment will confer kanamycin resistance on its host cell. Thus, the DNA sequences from mutant symbiotic loci can be cloned using direct genetic selection. This approach has been used to clone DNA from eight mutant symbiotic loci in *R. meliloti*. An agarose gel containing the *EcoRI* digests of several of these clones is shown in Fig. 9.6.

The availability of cloned mutant sequences from symbiotic loci makes several new avenues of experimentation available. For example, the *R. meliloti* sequences on these plasmids will be homologous with regions in wild type *R. meliloti* symbiotic genes. This homology can be exploited to obtain cloned wild type genes by screening shotgun gene banks of *R. meliloti* DNA. Such attempts are currently in progress. Wild type symbiotic genes could be useful for complementation studies with other types of symbiotic mutants. Also, their transfer into other species of *Rhizobium* might help elucidate the genetic basis for host specificity.

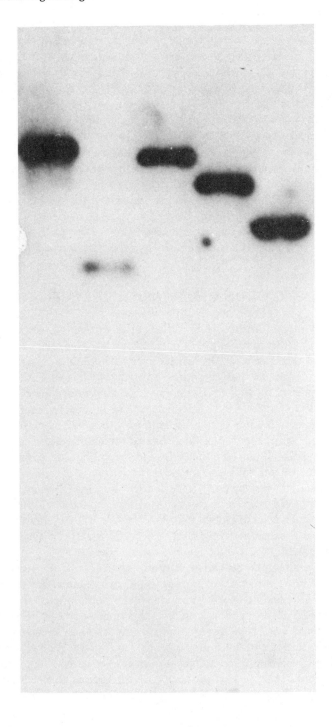

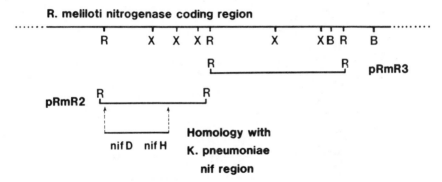

FIG. 9.4. Restriction map of the *R. meliloti* genomic region containing sequences homologous to the cloned *nif* structural genes of *K. pneumoniae*. pRmR2 and pRmR3 are recombinant plasmids containing adjacent *Eco*RI fragments from wild type *R. meliloti* ligated into the *Eco*RI site of pACYC184. The *R. meliloti* DNA insert in pRmR2 is 3.9 kb in length, and that in pRmR3 is 5.0 kb. Part of the insert in pRmR2 is homologous to the cloned *nif* D and *nif* H sequences of *K. pneumoniae*. The construction and analysis of these clones is described in Ruvkun and Ausubel (1981). R = *Eco*RI; X = *Xho*I; B = *Bam*HI.

In addition, the cloned mutant sequences now available can be used to determine whether the genes they represent are transcribed in free-living *Rhizobium* cells by examining ribonucleic acid (RNA) populations for sequence homology with the cloned DNA. Such an experiment could answer the question of whether these loci — identified because their inactivation leads to symbiotic failure — are uniquely expressed in the symbiosis or whether they are necessary for but not exclusive to the symbiotic process.

SUMMARY

We have examined the genetics of the bacterium *R. meliloti* as related to its symbiosis with alfalfa, which normally results in the development of nitro-

FIG. 9.3. Hybridization of radioactively labeled Tn5 DNA to digested DNA from Tn5-containing strains of *R. meliloti*. Each lane contains an *Eco*RI digest of total DNA from a symbiotic *R. meliloti* mutant obtained from the suicide plasmid mutagenesis and presumably containing a Tn5 insert. Fragments resulting from the *Eco*RI digestion were separated by gel electrophoresis (as in Fig. 9.6, lanes **a** and **b**) and transferred from the gel onto nitrocellulose filters by the method of Southern (1975). Tn5 DNA was obtained by preparing plasmid DNA that contained Tn5 transposed into ColE1 (Meade, unpubl.); the plasmid was radioactively labeled with ^{32}P by nick translation and hybridized with the nitrocellulose filter as described in Meade et al. (submitted).

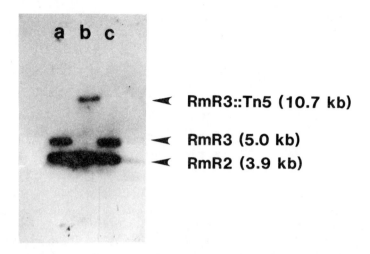

FIG. 9.5. Hybridization of pRmR2 and pRmR3 to *Eco*RI digests of *R. meliloti* DNA. (a) DNA from wild type *R. meliloti*, strain Rm1021 (no Tn5). (b) DNA from Fix⁻ mutant 1128, obtained from the Tn5 mutagenesis; insertion of Tn5 into the *Eco*RI fragment represented by pRmR3 has changed the mobility of the band that is homologous to the radioactively labeled pRmR3 in the hybridization probe. (c) Fix⁻ mutant 1080, another Tn5-induced symbiotic mutant. The insertion of Tn5 has occurred in a region outside of that represented in pRmR2 and pRmR3 because the mobilities of the fragments from that region are unchanged.

gen-fixing root nodules. Obtaining and analyzing genetic mutants of the bacterium that affect the symbiosis is made more difficult because the symbiotic phenotype itself cannot be monitored in culture; to overcome this difficulty, we have used transposon mutagenesis, which provides a physical and genetic marker at the mutated site. We have isolated a series of Tn5-induced mutants of *R. meliloti* that do not successfully interact with a host alfalfa plant to form a functioning symbiotic nodule. Using a combination of criteria, we have established for some of these mutants that the symbiotic defect is caused by the insertion of a single Tn5 element. These Tn5 insertions define symbiotic genes that, because they do not confer auxotrophy or any other phenotypic expression in free-living cells, probably contain specific information for nodulation and nitrogen fixation. Among the mutants are some that appear to block nodule formation at particular developmental stages. Such mutants should be useful for analyzing sequential, reciprocal interaction of host and bacterium during nodule formation.

The physical analysis performed on DNA from symbiotic mutants has confirmed the nature of the Tn5 insertion and made possible a test for inser-

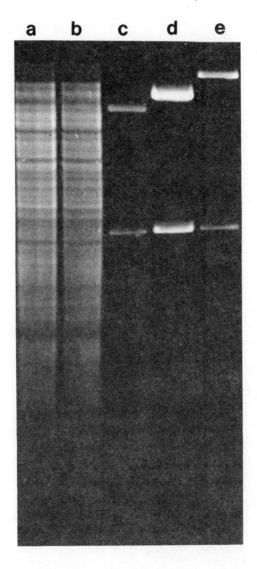

FIG. 9.6. Agarose gel electrophoresis. Samples were digested with *Eco*RI.
(a) *Eco*RI digest of wild type *R. meliloti* DNA. (b) *Eco*RI digest of DNA from
Tn5-containing strain 1107. (c) *Eco*RI digest of plasmid pRmR5, containing a segment of *R. meliloti* DNA from a *met* gene with Tn5 inserted into it (*met*::Tn5). Two
small, noncontiguous fragments were cloned into the plasmid along with the
met::Tn5 region. (d) Plasmid pRmSL5, containing DNA from the Tn5 mutated *fix*
gene in *R. meliloti* mutant Rm1088 (*fix*::Tn5). (e) pRmSL3, containing the *fix*::Tn5
fragment of Fix⁻ mutant Rm1083.

tions in the *nif* region of *R. meliloti*. Through recombinant DNA techniques we have cloned *R. meliloti* DNA fragments containing parts of *nod* and *fix* genes. Using these Tn5-containing clones as hybridization probes, it should now be possible to clone nonmutated restriction fragments from wild type *R. meliloti* and to use the cloned fragments in further genetic tests.

Genetics is an important tool for analyzing complex developmental processes such as symbiotic root nodule formation. Our work so far has concentrated only on the genetics of the bacterium, which is better defined genetically and more experimentally manipulable than the plant. A complete understanding of this system will require the application of genetic analysis to the plant partner as well as the bacterium. These genetic studies, together with biochemical studies of nodules, should help elucidate how individual gene products function during nodule formation and nitrogen fixation.

REFERENCES

Amarger, N. 1975. Efficience symbiotique de mutants spontanés de *Rhizobium leguminosarum* résistants à la streptomycine, spectinomycine ou kanamycine. *Comptes Rendus Acad. Sci.* (D) 280:1911–14.

Berg, D. E. 1977. Insertion and excision of the transposable kanamycin resistance element Tn5. In *DNA insertion elements plasmids and episomes*, ed. A. I. Bukhari, J. A. Shapiro, and S. L. Adhya, Cold Spring Harbor, N.Y.: Cold Spring Harbor Laboratory.

Bergersen, F. J., and Nutman, P. S. 1957. Symbiotic effectiveness in nodulated red clover. IV. The influence of the host factors i and ie upon nodule structure and cytology. *Heredity* 11:175–84.

Beringer, J. E.; Beynon, J. L.; Buchanan-Wollaston, A. V.; and Johnston, A. W. B. 1978. Transfer of the drug-resistance transposon Tn5 to *Rhizobium*. *Nature* (London) 276:633–34.

Beringer, J. E.; Johnston, A. W. B.; and Wells, B. 1977. The isolation of conditional ineffective mutants of *Rhizobium leguminosarum*. *J. Gen. Microbiol.* 98: 339–43.

Cannon, F. C.; Riedel, G. E.; and Ausubel, F. M. 1979. Overlapping sequences of *Klebsiella pneumoniae nif* DNA cloned and characterized. *Molec. Gen. Genet.* 174:59–66.

Dénarié, J.; Truchet, G.; and Bergerson, B. 1976. Effects of some mutations on symbiotic properties of *Rhizobium*. In *Symbiotic nitrogen fixation in Plants*, ed. P. S. Nutman, pp. 47–61. London: Cambridge University Press.

Fåhraeus, G. 1957. The infection of clover root hairs by nodule baceria studied by a simple glass slide technique. *J. Gen. Microbiol.* 16:374–81.

Hendry, G. S., and Jordan, D. C. 1969. Ineffectiveness of viomycin-resistant mutants of *Rhizobium meliloti*. *Can. J. Microbiol.* 15:671–75.

Kleckner, N. 1977. Translocatable elements in prokaryotes. *Cell* 11:11–23.

Kondorosi, A.; Kiss, G. B.; Forrai, T.; Vincze, E.; and Banfalvi, Z. 1977. Circular map of *Rhizobium meliloti* chromosome. *Nature* (London) 268:525–27.

Kondorosi, A.; Vincze, E.; Johnston, A. W. B.; and Beringer, J. E. 1980. A comparison of three *Rhizobium* linkage maps. *Molec. Gen. Genet.* 178:403–8.

Legocki, R. P., and Verma, D. P. S. 1980. Identification of "nodule-specific" host proteins (nodulins) involved in the development of *Rhizobium*-legume symbiosis. *Cell* 20:153–64.

Libbenga, K. R., and Bogers, R. J. 1974. Root-nodule morphogenesis. In *The biology of nitrogen fixation*, ed. A. Quispel, pp. 430–72. Amsterdam: North-Holland.

Long, S. R.; Meade, H. M.; Ruvkun, G. B.; Brown, S. E.; and Ausubel, F. M. Submitted. Molecular and genetic analysis of Tn5-induced symbiotic mutants of *Rhizobium meliloti*.

Maier, R. J., and Brill, W. J. 1976. Ineffective and non-nodulating mutant strains of *Rhizobium japonicum*. *J. Bacteriol.* 127:763–69.

Meade, H. M.; Long, S. R.; Ruvkun, G. B.; Brown, S.; and Ausubel, F. M. Submitted. Isolation of symbiotic and auxotrophic mutants of *Rhizobium meliloti* using transposon Tn5 mutagenesis.

Meade, H. M., and Singer, E. E. 1977. Genetic mapping of *Rhizobium meliloti*. *Proc. Nat. Acad. Sci. (USA)* 74:2076–78.

Mosse, B. 1964. Electron-microscope studies of nodule development in some clover species. *J. Gen. Microbiol.* 36:49–66.

Nutman, P. S. 1971. The modified Fåhraeus slide technique. In *A manual for the practical study of root-nodule bacteria*, ed. J. M. Vincent, pp. 144–45. Oxford and London: Blackwell Scientific.

Pankhurst, C. E., and Schwinghamer, E. A. 1974. Adenine requirement for nodulation of pea by an auxotrophic mutant of *Rhizobium leguminosarum*. *Arch. Microbiol.* 100:219–38.

Ruvkun, G. B., and Ausubel, F. M. 1981. A general method for site-directed mutagenesis in prokaryotes. *Nature* (London) 289:85–88.

Schwinghamer, E. A. 1968. Loss of effectiveness and infectivity in mutants of *Rhizobium* resistant to metabolic inhibitors. *Can. J. Microbiol.* 14:355–67.

——— . 1977. Genetic aspects of nodulation and dinitrogen fixation by legumes: the microsymbiont. In *A treatise on dinitrogen fixation, section III: biology*, ed. R. W. F. Hardy and W. S. Silver, pp. 577–622. New York: John Wiley Sons.

Southern, E. M. 1975. Detection of specific sequences among DNA fragments separated by gel electrophoresis. *J. Mol. Biol.* 98:503–17.

Truchet, M.; Michel, M.; and Dénarié, J. 1980. Sequential analysis of the organogenesis of lucerne (*Medicago sativa*) root nodules using symbiotically-defective mutants of *Rhizobium meliloti*. *Differentiation* 16:163–72.

Van Vliet, F.; Silva, B.; Van Montague, M.; and Schell, J. 1978. Transfer of RP4::Mu plasmids to *Agrobacterium tumefaciens*. *Plasmid* 1:446–55.

Vincent, J. M. 1974. Root-nodule symbioses with *Rhizobium*. In *The biology of nitrogen fixation*, ed. A. Quispel, pp. 266–341. Amsterdam: North-Holland.

——— . 1980. Factors controlling the legume-*Rhizobium* symbiosis. In *Nitrogen fixation. Volume II: symbiotic associations and cyanobacteria*, ed. W. E. Newton and W. H. Orme-Johnson. Baltimore: University Park Press.

10

A NEW BROAD-HOST-RANGE DNA CLONING VECTOR FOR USE WITH *RHIZOBIUM* AND OTHER GRAM-NEGATIVE BACTERIA

Gary Ditta

Numerous deoxyribonucleic acid (DNA) cloning vectors are available for use in *Escherichia coli*, including a variety of plasmid (Bolivar, Rodriguez, Betlach, and Boyer 1977; Bolivar, Rodriguez, Greene, Betlach, Heyneker, Boyer, Crosa, and Falkow 1977; Chang and Cohen 1978; Kahn et al. 1979), phage (Blattner et al. 1977), and plasmid-phage hybrid vectors (Collins and Bruening 1978; Collins and Hohn 1978). Unfortunately, these vectors possess a very limited host range; they cannot be used to transfer cloned DNA to many bacteria that are of significant medical and agricultural importance. To help rectify this situation, a plasmid cloning system with broad-host-range transfer and replication potential has been developed from the drug resistance plasmid RK2. In this system DNA fragments are first cloned into *E. coli* and subsequently transferred at high frequency to other gram-negative bacteria.

BINARY VEHICLE SYSTEM

RK2 is a self-replicating, circular DNA molecule of approximately 56 kilobase pairs (kbp), which confers resistance to the antibiotics ampicillin, kanamycin, and tetracycline (Datta and Hedges 1972). It is very similar, if not identical, to other plasmids having the designations RP1, RP4, and R68 (Burkardt, Riess, and Puhler 1979). RK2 is a member of the incompatibility group P-1 and, like other members of this incompatibility group, has the ability to transfer itself via conjugation at high frequency to a wide variety of gram-negative bacteria. It is thus a logical choice for further development as a broad-host-range cloning vehicle.

The essential features one would expect in a broad-host-range cloning vector are:

The ability to insert exogenous DNA into the vector in vitro using recombinant DNA technology,

The ability to select for cells carrying the vector, and

The ability of the vector to be introduced and stably maintained in a wide variety of bacterial hosts.

While native RK2 DNA has been used directly as a recombinant cloning vector, its large size is a serious impediment to routine use. It was possible to reduce the size of RK2 and at the same time increase the overall biological containability by designing the resulting plasmid to be part of the two-plasmid, binary vehicle system shown diagrammatically in Fig. 10.1. The possible development of such a system was first suggested some time ago (Meyer, Figurski, and Helinski 1977). In this system the RK2 broad-host-range transfer genes (*tra*), which are capable of acting in *trans* on a suitable mobilizable plasmid, were separated from the broad-host-range replication capability (Rep) of RK2 by placing the genes on different plasmids. The vector is a deletion derivative of RK2 called pRK290. It is com-

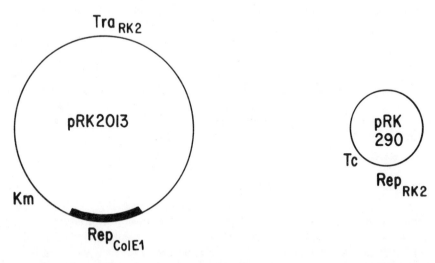

FIG. 10.1. Binary vehicle system. pRK290 is the cloning vector. pRK2013 is a helper plasmid essential for conjugal transfer of pRK290. Rep and Tra refer to generalized replication and conjugal transfer capability, respectively, and are not intended to designate specific genes.

posed entirely of RK2 DNA, codes for tetracycline resistance, and is mobilizable but nonconjugative. The helper plasmid, pRK2013, is a hybrid composed of the narrow-host-range plasmid ColE1 and the entire complement of RK2 transfer genes. It confers resistance to kanamycin and is capable of mobilizing pRK290.

To use this system DNA is inserted into pRK290 in vitro and is transformed into *E. coli* using standard procedures. Once established in *E. coli*, the hybrid DNA can be conjugally transferred to other gram-negative bacteria following the introduction of pRK2013 into the *E. coli* host. Cloning into *E. coli* as an intermediate host has several advantages over attempting to transform directly into the intended recipient. First, the development of a suitable transformation procedure for certain bacteria can be a difficult or impossible task. Second, in *E. coli* one has ready access to a wide variety of techniques for rapid plasmid DNA analysis. Third, in *E. coli* a heterologous DNA background exists for the cloned DNA. This is valuable since the cloned DNA may show recombinational instability in its host or origin, and it allows radioactive probes to be used to identify particular cloned sequences. In comparison with some bacteria the rapid growth rate of *E. coli* can also be an important consideration.

The construction of the helper plasmid pRK2013 has been reported previously in connection with a series of experiments investigating RK2 replication (Figurski and Helinski 1979). The cloning vector pRK290 has recently been derived from RK2 as part of a construction designed to delete as much DNA as possible while retaining mobilizability and a single drug resistance marker.

Knowledge of the regions of RK2 that are essential for replication, transfer, and mobilization were necessary for this construction. These regions are shown in Fig. 10.2. *trf*A and *trf*B refer to *trans*-acting replication functions capable of activating the origin of replication, ori_{RK2} (Thomas, Meyer, and Helinski 1979). *rlx* is a *cis*-acting site necessary for mobilizability (Guiney and Helinski 1979). *tra* regions are necessary for conjugal transfer (Barth, Grinter, and Bradley 1978). In constructing pRK290 it was necessary to retain functions on diverse segments of the molecule while bringing about a considerable reduction in size. Details of the steps used in this construction are presented elsewhere (Ditta et al. 1980). A restriction enzyme map of pRK290 is shown in Fig. 10.3. It is 20 kbp in size and has two single restriction enzyme sites that are available for cloning — a *Bgl*II site and an *Eco*RI site. Despite its relatively large size, pRK290 lacks cleavage sites for many commonly used restriction enzymes, including *Kpn*I, *Pst*I, *Bam*HI, *Hind*III, *Hpa*I, and *Xho*I.

pRK290 is efficiently mobilized by pRK2013 into a variety of gram-negative bacteria. We have specifically tested and found high rates of transfer to *Klebsiella pneumoniae*, *Serratia marcescens*, *Pseudomonas aerugi-*

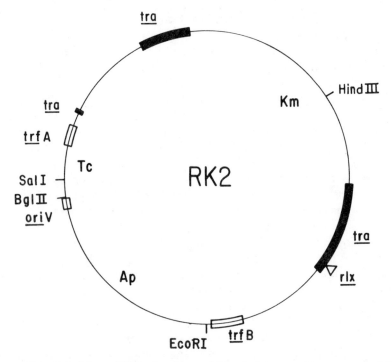

FIG. 10.2. Map of RK2. **Ap, Tc,** and **Km** designate resistance to ampicillin, tetracycline, and kanamycin, respectively. *trf* A and *trf* B refer to *trans*-acting replication functions (Thomas, Meyer, and Helinski 1979). *ori* V is the origin of replication as determined by electron microscopy (Meyer and Helinski 1977). *rlx* refers to the relaxation complex site (Guiney and Helinski 1979). *tra* refers to genes required for conjugal transfer (Barth, Grinter, and Bradley 1978).

nosa, Acinetobacter calcoaceticus, and *Rhizobium meliloti*. Conjugal transfer data for *E. coli* and for *R. meliloti*, the root nodule bacteria capable of infecting alfalfa, are presented in Table 10.1. The *E. coli* host strain HB101, a recombination-deficient mutant, was used in order to prevent recombination between helper and vector, which share large regions of homology. HB101 is also an appropriate strain for cloning experiments, since it lacks a restriction system that might otherwise cause the loss of cloned inserts. Bacterial matings were performed by combining washed cultures of donors and recipients, collecting the mixed cells on a nitrocellulose filter, and incubating the filter for several hours on a nutrient agar surface. The cells were then resuspended and plated on selective media.

It can be seen that the tetracycline-resistant vector pRK290 is incapable of self-transfer to either *E. coli* or *R. meliloti* (lines 2 and 5) but that in combination with pRK2013 it shows a very high rate of vector transfer to

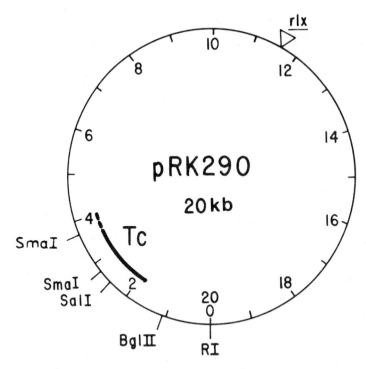

FIG. 10.3. Physical map of pRK290. Coordinates are in kbp. *rlx* is the relaxation complex site (Ditta et al. 1980).

both of these hosts (lines 3 and 6). This high rate of transfer could be particularly valuable in situations where cloned DNA must be transferred into a restriction-competent cell line, since a loss of several orders of magnitude in the transfer frequency could easily be tolerated. Also, for maximal biological containment, as well as ease of manipulation, such matings can be done as triparental crosses, where helper and vector are in separate strains that are mixed together with the recipient. Under these conditions the self-transmissible helper moves into the vector-containing strain and subsequently promotes mobilization of pRK290 into the recipient. Triparental mating frequencies are comparable to those of biparental crosses (lines 6 and 7).

The behavior of the helper plasmid in this system is quite interesting. Although pRK2013 alone transfers well to *E. coli* (line 1), it transfers poorly to *Rhizobium* (line 4). This is presumably a reflection of the relatively narrow host range of ColE1. As part of the binary plasmid system (line 6), however, pRK2013 transfers at a much higher frequency to *Rhizobium*, approximately 1 percent of that observed for pRK290. The most logical ex-

TABLE 10.1————————
Plasmid Transfer Frequencies

Donor	Recipient			$\dfrac{Km^R\ Conjugants}{Recipients}$	$\dfrac{Tc^R\ Conjugants}{Recipients}$
E. coli					
HB101(pRK2013)	*E. coli*	HB101	*rif*	8.5×10^{-1}	n.a.
HB101(pRK290)	*E. coli*	HB101	*rif*	n.a.	0
HB101(pRK2013, pRK290)	*E. coli*	HB101	*rif*	8.2×10^{-1}	4.0×10^{-1}
HB101(pRK2013)	*R. meliloti*	104B5	*nal*	1.7×10^{-7}	n.a.
HB101(pRK290)	*R. meliloti*	104B5	*nal*	n.a.	0
HB101(pRK2013, pRK290)	*R. meliloti*	104B5	*nal*	8.4×10^{-4}	4.6×10^{-2}
HB101(pRK2013) + HB101(pRK290)	*R. meliloti*	104B5	*nal*	5.6×10^{-4}	8.3×10^{-2}

n.a. = not available
Km^R = kanamycin resistant
Tc^R = tetracycline resistant
Source: Data taken from Ditta et al. 1980.

planation is that this represents recombinational rescue of pRK2013 in the *rec*⁺ *Rhizobium* by pRK290. All kanamycin-resistant conjugants in this case are seen to be tetracycline resistant. pRK2013 will of course cotransfer with pRK290 to those hosts where pRK2013 alone transfers. Such is obviously the case for *E. coli*. Under these circumstances it is quite easy to segregate clones lacking pRK2013 since both vector and helper display mutual incompatibility when present in the same cell. A simple restreaking on tetracycline is usually enough to generate kanamycin-sensitive clones from kanamycin-resistant exconjugants.

Rhizobium DNA has been cloned into pRK290 as inserts ranging in size up to 50 kbp. Examples of cloned DNAs can be seen in Fig. 10.4. For each clone the inserted DNA has been excised and displayed in the form of its component *Bgl*II fragments. Constructions were performed by ligating partially digested *Rhizobium* DNA to pRK290 that had been cleaved with *Bgl*II and subsequently treated with bacterial alkaline phosphatase. The latter treatment prevented covalent recircularization of the vector by DNA ligase and ensured that the majority of transformants represented molecules carrying cloned DNA inserts. The ability to isolate large inserts can be very important. In constructing a gene bank, for instance, the larger the insert, the smaller the number of clones necessary to provide complete sequence

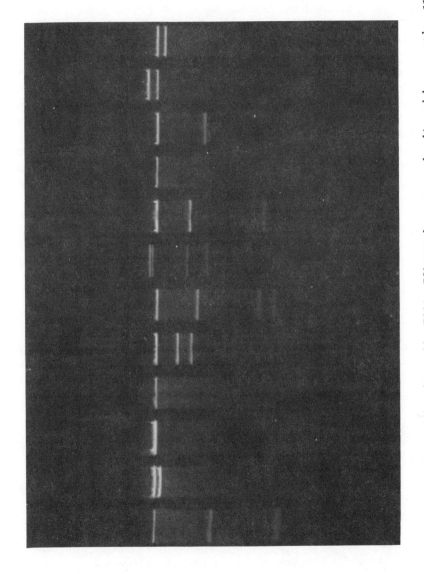

FIG. 10.4. *Bgl*II digests of cloned *R. meliloti* DNA. pRK290 is the uppermost band in each lane except lane 11. Lane 7 is a set of λ DNA-size standards.

representation. Large clones also increase the probability that functional gene clusters will be preserved.

In certain kinds of experiments the stability of pRK290 may be an important consideration. Although tetracycline can normally be used to ensure maintenance of the vector, situations exist where direct antibiotic selection is impossible or undesirable. An example of this would be the inoculation of legumes by rhizobia carrying cloned DNA. To test the stability of pRK290 in *Rhizobium*, clones were allowed to develop from individual cells of *R. meliloti* on nonselective agar plates. A large number of these colonies, each representing approximately 25 cell divisions, were dispersed in broth by vortexing and were again plated on nonselective medium. Individual colonies from these plates were then picked onto selective versus nonselective media. The average loss per generation in *E. coli* and in *R. meliloti* for pRK290 with and without several random cloned inserts is given in Table 10.2. From these data it can be seen that the vector is quite stable in both bacterial hosts, though not completely so in *Rhizobium*, and that DNA insertion per se at either the *EcoRI* or *BglII* site does not significantly alter stability. The somewhat lower stability of plasmid pRK290B9 would suggest that individual cloned fragments may have different effects on overall stability. In general, however, the low rate of plasmid loss observed here should not be a problem except under the most demanding of circumstances.

TABLE 10.2———
Plasmid Stability

Strain	Generations without Selection	Percentage Loss per Generation
E. coli		
HB101(pRK290)	25	0.0
HB101(pRK290 R2)	25	0.3
HB101(pRK290 R9)	25	0.1
HB101(pRK290 B3)	n.a.	0.1
HB101(pRK290 B9)	25	3.0
R. meliloti		
2011(pRK290)	27	0.2
2011(pRK290 R2)	26	0.4
2011(pRK290 R9)	26	0.3
2011(pRK290 B3)	26	0.3
2011(pRK290 B9)	26	1.6

n.a. = not available
Note: R2 and R9 are cloned *EcoRI* fragments; B3 and B9 are cloned *BglII* fragments.
Source: Compiled by the author.

SUMMARY

A broad-host-range cloning vector has been developed from RK2 that offers the advantages of reduced size and increased biological containability. It can be mobilized at high frequency from *E. coli* to other gram-negative bacteria in the presence of a helper plasmid but is otherwise non-self-transmissible. *R. meliloti* DNA has been successfully cloned with this vector.

REFERENCES

Barth, P. T.; Grinter, H. J.; and Bradley, D. E. 1978. Conjugal transfer system of plasmid RP4: analysis by transposon 7 insertions. *J. Bacteriol.* 133:43–52.

Blattner, F. R.; Williams, B. G.; Blechl, A. E.; Thompson, K. D.; Faber, H. E.; Furlong, L. A.; Grunwald, D. J.; Kiefer, D. O.; Moore, D. D.; Schumm, J. W.; Sheldon, E. L.; and Smithies, O. 1977. Charon phages: safer derivatives of bacteriophage λ for DNA cloning. *Science* 196:161–69.

Bolivar, F.; Rodriguez, R. L.; Betlach, M. C.; and Boyer, H. W. 1977. Construction and characterization of new cloning vehicles. I. Ampicillin-resistant derivatives of the plasmid pMB9. *Gene* 2:79–93.

Bolivar, F.; Rodriguez, R. L.; Greene, P. J.; Betlach, M C.; Heyneker, H. L.; Boyer, H. W.; Crosa, J. H.; and Falkow, S. 1977. Construction and characterization of new cloning vehicles. II. A multipurpose cloning system. *Gene* 2:95–113.

Burkardt, H. -J.; Riess, G.; and Puhler, A. 1979. Relationship of Group P1 plasmids revealed by heteroduplex experiments: RP1, RP4, R68, and RK2 are identical. *J. Gen. Microbiol.* 114:341–48.

Chang, A. C. Y., and Cohen, S. N. 1978. Construction and characterization of amplifiable multicopy DNA cloning vehicles derived from the P15A cryptic mini-plasmid. *J. Bacteriol.* 134:1141–56.

Collins, J., and Bruening, H. J. 1978. Plasmids usable as gene-cloning vectors in an *in vitro* packaging by coliphage: cosmids. *Gene* 4:85–107.

Collins, J., and Hohn, B. 1978. Cosmids: a type of plasmid gene-cloning vector that is packageable *in vitro* in bacteriophage heads. *Proc. Nat. Acad. Sci. (USA)* 75:4242–46.

Datta, N., and Hedges, B. W. 1972. Host ranges of R Factors. *J. Gen. Microbiol.* 70:453–60.

Ditta, G.; Stanfield, S.; Corbin, D.; and Helinski, D. R. 1980. Broad host range cloning system for gram-negative bacteria: construction of a gene bank of *Rhizobium meliloti*. *Proc. Nat. Acad. Sci. (USA)* 77:7347–51.

Figurski, D., and Helinski, D. R. 1979. Replication of an origin containing derivative of plasmid RK2 dependent on a plasmid function provided in *trans*. *Proc. Nat. Acad. Sci. (USA)* 76:1648–52.

Guiney, D. G., and Helinski, D. R. 1979. The DNA-protein relaxation complex of plasmid RK2: location of the site-specific nick in the region of the proposed origin of transfer. *Mol. Gen. Genet.* 176:183–89.

Kahn, M.; Kolter, R.; Thomas, C.; Figurski, D.; Meyer, R.; Remaut, E.; and Helinski, D. R. 1979. Plasmid-cloning vehicles derived from plasmids ColE1, F, R6K, and RK2. In *Methods in enzymology*, ed. S. P. Collowick and N. O. Kaplan, vol. 68, pp. 268–80. New York: Academic Press. (Edited by R. Wu)

Meyer, R. J.; Figurski, D.; and Helinski, D. R. 1977. Properties of the plasmid RK2 as a cloning vehicle. In *DNA insertion elements, plasmids and episomes*, ed. A. I. Bukhari, J. A. Shapiro, and S. L. Adhya, pp. 559–66. Cold Spring Harbor, N.Y.: Cold Spring Harbor Laboratories.

Meyer, R. J., and Helinski, D. R. 1977. Unidirectional replication of the P-group plasmid RK2. *Biochim. Biophys. Acta* 478:108–13.

Thomas, C. M.; Meyer, R.; and Helinski, D. R. 1979. Regions of broad-host-range plasmid RK2 which are essential for replication and maintenance. *J. Bacteriol.* 141:213–22.

11
PLASMID RSF1010 AS A CLONING VECTOR IN *PSEUDOMONAS*

Kenji Nagahari

INTRODUCTION

Since the cloning technique was first successfully developed by Cohen et al. (1973), *Escherichia coli* has served as the host in the cloning of genes from various organisms, prokaryotes as well as eukaryotes. DNA cloning systems have also been developed recently in other industrially important species such as *Bacillus subtilis* (Gryczan, Shivakumar, and Dubnau 1980; Tanaka and Sakaguchi 1978) and *Saccharomyces cerevisiae* (Beggs 1978; Broach, Strathern, and Hicks 1979; see also Chapter 13.).

Bacteria belonging to the genus *Pseudomonas* play important roles in natural and agricultural environments. They are widely distributed in nature as soil saprophytes, plant epiphytes, or pathogens and are often capable of using diverse organic compounds as sole carbon sources. Chakrabarty et al. (1975) reported the successful transformation of *Pseudomonas putida* with the broad-host-range plasmid RP1. We (Nagahari et al. 1977) previously constructed, by cloning in *E. coli*, four hybrid plasmids (pSC101-*trp*, RSF1010-*trp*, RSF2124-*trp*, and RP4-*trp*), each of which consisted of a vector plasmid and a 10.8 Mdal *Eco*RI fragment containing the *E. coli* tryptophan operon. These hybrid plasmids were stably maintained in *E. coli trpB* cells and permitted expression of the *trp* genes—that is, converted the Trp⁻ phenotype into Trp⁺ (tryptophan prototrophic). To investigate the expression and regulation of the *E. coli trp* operon in genetically unrelated bacteria, such as *Pseudomonas* or *Rhizobium* species, the RP4-*trp*

This work was performed while the author was in the Laboratory of Microbiological Chemistry, Mitsubishi-Kasei Institute of Life Sciences. I am grateful to Dr. K. Sakaguchi for his encouragement and to Ms. T. Koshikawa for technical assistance.

plasmid was conjugatively transferred from *E. coli* to tryptophan-deficient mutants of *P. aeruginosa* and *Rhizobium leguminosarum*. Trp⁺ conjugants appeared on the plates lacking tryptophan and constitutive expression of tryptophan-synthesizing enzymes encoded by *E. coli trp* operon were observed in both species (Nagahari, Koshikawa, and Sakaguchi 1979; Nagahari, Sano, and Sakaguchi 1977). The expression of the *E. coli trp* operon in *P. aeruginosa* made it possible to investigate whether the other three hybrid plasmids (pSC101-*trp*, RSF1010-*trp*, and RSF2124-*trp*) were able to replicate in and be maintained by *P. aeruginosa trp* cells, because the Trp⁺ phenotype of *E. coli trp* operon on the 10.8 Mdal *Eco*RI fragment could be used as a selective marker. It was in the course of these experiments that RSF1010 was first found to be capable of replication in *Pseudomonas* species.

MATERIALS AND METHODS

Bacterial Strains and Plasmids

E. coli J5, which harbors the RSF1010 plasmid, and *E. coli* C600*hsdR hsdM trp* containing each of the four *trp* hybrid plasmids have been described (Nagahari et al. 1977). *P. aeruginosa* M12 (Nagahari, Sano, and Sakaguchi 1977), *P. putida* ATCC12633 (wild type), and *P. putida trpB* (Nagahari, Koshikawa, and Sakaguchi 1979) were used as recipient cells for transformation.

Media

Media used for the growth of *P. aeruginosa* or *P. putida* were those described previously (Nagahari and Sakaguchi 1978; Nagahari, Sano, and Sakaguchi 1977).

Experimental Procedure

Transformation of *P. aeruginosa* and *P. putida* cells were performed according to the method of Sano and Kageyama (1977) and that of Chakrabarty et al. (1975), respectively. Plasmid isolation, digestion of deoxyribonucleic acid (DNA) with restriction endonuclease, and electrophoresis in agarose gel have been described elsewhere (Nagahari 1978).

RESULTS

Transformation of *P. aeruginosa trp* Cells with Hybrid Plasmids Containing the *E. coli trp* Operon

P. aeruginosa trp cells exposed to $CaCl_2$ pretreatment were transformed with DNA of the four hybrid plasmids (pSC101-*trp*, RSF1010-*trp*, RSF2124-*trp*, and RP4-*trp*). Transformation with either RSF1010-*trp* or RP4-*trp* plasmid yielded colonies that were tryptophan independent. All the transformants with RSF1010-*trp* selected against Trp⁺ were also Sm^R (streptomycin resistant), which was conferred by RSF1010 plasmid, suggesting that the phenotypic change of the cells was caused by the presence of RSF1010-*trp* plasmid in the cells. From the transformants various plasmid mutants were isolated that lacked parts of the original plasmid unnecessary for the expression of Trp⁺ and the maintenance of the plasmid. The structures of these plasmids were determined by their digestion patterns with several restriction endonucleases (Nagahari 1978). The portions deleted from the original plasmid were found to have one end located between two *Pst*I sites (0.5 Mdal) of RSF1010, while the other end was in varied locations within the 10.8 Mdal *Eco*RI fragment.

The mechanism of deletion formation in these plasmids is not yet clear. Deletion plasmids were also isolated from transformants previously heated at 42°C, a treatment that is used to inactivate the restriction system of *P. aeruginosa* host cells (Holloway 1965). RSF1010-*trp* was also introduced into *P. putida trpB* by transformation. In this case, however, no deletion plasmids were detected in the transformants.

The experimental results described above suggested that the RSF1010 plasmid could be used as a cloning vector in *Pseudomonas* species. Therefore, the plasmid was further characterized.

Characterization of RSF1010 in *P. putida*

P. putida ATCC12633 cells were transformed into Sm^R with RSF1010 plasmid DNA (Nagahari and Sakaguchi 1978). The transformation efficiency was about 10^{-7} per recipient cell using the RSF1010 plasmid DNA purified from *E. coli* J5 cells. However, with RSF1010 plasmid derived from *P. putida* ATCC12633 transformants, transformation efficiency was greater than 10^{-5} per recipient cell, yielding 2.8×10^5 transformants per 1 µg of RSF1010 DNA. Shaking for 4 hours in liquid medium in the absence of streptomycin was found necessary for full expression of the Sm^R phenotype in *P. putida* cells after transformation.

The minimal inhibitory concentration of streptomycin for the transformed cell was 1,000-fold higher than that for the parent strain (10 $\mu g/ml$), suggesting a high copy number of RSF1010 plasmid in *P. putida* cells. Therefore, the copy number of RSF1010 plasmid was determined by cesium chloride-ethidium bromide (CsCl-EtBr) density gradient centrifugation of the whole lysate of the transformed cells labeled with [^{3}H]-adenosine. By this method only the supercoiled DNA (lower band) can be estimated. The molecular weight of the *P. putida* chromosomal DNA was assumed to be 4×10^9 (Lethbak, Christiansen, and Stenderup 1970) and that of RSF1010 5.5×10^6 (Nagahari et al. 1977), giving a copy number of 43 per chromosome. However, plasmid molecules present as open-circle DNA can not be calculated by this method. The upper-band DNA in the CsCl-EtBr gradient was, therefore, digested with *Eco*RI restriction endonuclease, and the density of the specific band in the agarose gel arising from the plasmid DNA against the background derived from the digested chromosomal DNA was measured with a densitometer. From these measurements an additional 123 copies of RSF1010 were estimated. Based on the above results, the copy number of RSF1010 in *P. putida* appears to be over 150 per chromosome.

RSF1010 was more stable in *P. putida* than in *E. coli*. After one generation in liquid culture lacking streptomycin, 0.2 percent of *P. putida* cells lost the Sm^R phenotype as opposed to 1.2 percent in *E. coli*.

Expression of the *E. coli* Tryptophan Operon in *P. putida*

To study the mechanism of expression of *E. coli trp* operon in *P. putida*, many RSF1010-*trp* derivatives were constructed in vitro using *Pst*I or *Hind*III restriction endonuclease and T4 DNA ligase, followed by transformation of *E. coli*. These plasmids were subsequently purifed by CsCl-EtBr density gradient centrifugation, and their molecular maps were constructed (Fig. 11.1). Plasmid pKNT1 has lost three *Pst*I fragments not essential for the maintenance and the phenotypic expression (Trp$^+$, Sm^R) of the RSF1010-*trp* plasmid. pKNT3 and pKNT4, derived from pKNT1, had lost a 2.8 Mdal *Hind*III fragment on which the leftward promoter (P$_L$) and the *E. coli trp* operon promoter (P$_T$) are located but still retained a 1.8 Mdal *Hind*III fragment containing an internal promoter of the *E. coli trp* operon.

Table 11.1 shows the activities of two of the five enzymes involved in tryptophan synthesis found in the crude extracts of *P. putida* cells carrying each of the above plasmids. The crude extracts of the cells carrying RSF1010-*trp* or pKNT1, which contain an intact *E. coli trp* operon, had similar levels of anthranilate synthetase (ASase, product of *trpE*) activity but had much higher levels of tryptophan synthetase (TSaseβ, product of *trpB*) activity compared with those of the wild type cells. In contrast, cells

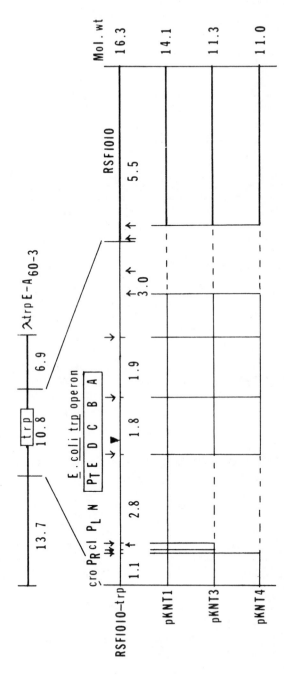

FIG. 11.1. Restriction maps of plasmid DNAs. Each plasmid is represented as a linear form whose left and right termini form an *Eco*RI site. Bold lines represent the RSF1010 plasmid DNA. (↓) depicts the cleavage site with *Hind*III; (↑) with *Pst*I; (▼) position of the internal promoter of the *E. coli* tryptophan operon. Size (Mdal) of DNA fragments generated by the double digestion of RSF1010-*trp* plasmid with *Eco*RI and *Hind*III are indicated below the line. E, D, C, B, and A: structural genes of the *E. coli* tryptophan operon; P_T: *trp* operon promoter; P_R: rightward promoter; P_L: λ leftward promoter.

159

TABLE 11.1——————————————
Specific Activities of Tryptophan-Synthesizing Enzymes in *P. putida* *trpB* Cells Carrying Derivatives of RSF1010-*trp* Hybrid Plasmid

Strain	Asase	TSaseβ
P. putida trpB		
RSF1010-*trp*	0.23	14.6
pKNT1	0.15	14.6
pKNT3	0.03	11.8
pKNT4	0.05	12.9
P. putida ATCC12633	0.40	1.4

Notes: A *unit of enzyme activity* is defined as the utilization or production of 1 nmole of substrate or product per minute at 37°C. Specific activity is given here as units per milligram of protein. Cells were cultured in Vogel and Bonner (1956) minimal medium.

Source: Compiled by the author.

carrying pKNT3 or pKNT4 had much less ASase activity but still retained high TSaseβ activity, despite the loss of P_L and P_T promoters (see Fig. 11.1). These results, together with other findings, suggest that the internal promoter of *E. coli trp* operon, from which low constitutive transcription occurs in *E. coli*, is stil functional in *P. putida* cells. The reason for the lower ASase activity in the cells carrying pKNT3 or pKNT4 compared with the wild type strain is not clear because *P. putida trpB* cells used as recipient cells should have an intact *trpE* gene on the chromosomal DNA.

Structure of the RSF1010 Plasmid and the Construction of the RSF1010-pBR322 Hybrid Plasmid

RSF1010 plasmid was a useful vector for cloning *Eco*RI fragments because it has a single unique *Eco*RI site into which DNA fragments can be inserted without abolishing the expression of Sm^R, the marker used in selection of transformants, and the ability to maintain itself (Nagahari et al. 1977). The mass of RSF1010 is 5.5 Mdal. It has, in addition to the *Eco*RI site, one single *Hpa*I site and two *Pst*I sites. These sites are located very close to the *Eco*RI site. As shown in Fig. 11.1, RSF1010-*trp* derivatives that lost the *Pst*I fragment of RSF1010 were still functional. These observations suggested that a DNA fragment could be inserted at or between the *Pst*I sites. Furthermore, RSF1010 derivatives lacking both *Pst*I sites were also obtained by in vivo deletion in *P. aeruginosa* (Nagahari 1978).

To obtain a more versatile and useful vector, an RSF1010-pBR322 hybrid plasmid was constructed in *E. coli* through ligation of the two linearized plasmids cleaved at their *Pst*I sites. As expected the newly derived plasmid conferred on the *E. coli* host a Sm[R], Ap[S] (ampicillin-sensitive), and Tc[R] (tetracycline-resistant) phenotype (insertion at the *Pst*I site of pBR322 causes the inactivation of the ampicillin resistance gene [Bolivar et al. 1977]). The restriction sites for *Bam*HI, *Sal*I and *Hind*III present in the Tc[R] gene on the pBR322 portion of RSF1010-pBR322 can be used for the insertion of DNA fragments, as RSF1010 has no such sites.

DISCUSSION

The *E. coli trp* operon was reported capable of expression in many gram-negative bacteria (Manson and Yanofsky 1976). In other experiments we successfully transferred the RP4-*trp* hybrid plasmid into *R. leguminosarum trp* cells (Nagahari, Koshikawa, and Sakaguchi 1979), converting their phenotype from Trp⁻ to Trp⁺. Therefore, if transformation systems are developed in gram-negative bacteria, the 10.8 Mdal *Eco*RI fragment carrying the *E. coli trp* operon would be useful in the development of a molecular cloning system in diverse species, as the tryptophan gene can be used as a selective marker.

RSF1010 has several advantages over RP4 as cloning vector in *Pseudomonas* species. It is nonconjugative and has a lower molecular weight and a higher copy number than RP4. The plasmid is therefore much more suitable for cloning experiments than RP4.

An RSF1010-pBR322 hybrid plasmid similar to the one described here was also constructed by Bagdasarian et al. (1979). These investigators were unable to introduce the plasmid into *Pseudomonas* cells. However, Panopoulos (Chapter 12) has shown that RSF1010-pBR322 hybrids can be introduced in phytopathogenic *Pseudomonas*. Bagdasarian et al. (1979) and Panopoulos (Chapter 12) also constructed various other derivatives of RSF1010 plasmid carrying other antibiotic resistance determinants. Recently, Gantotti, Patil, and Mandel (1979) reported the transformation of a phytopathogenic pseudomonad, *P. phaseolicola*, with RSF1010. The host range of RSF1010 among phytopathogenic *Pseudomonas* is very broad (Panopoulos, Chapter 12). Thus, RSF1010 is expected to be a strong candidate as a cloning vector in *Pseudomonas* species.

REFERENCES

Bagdasarian, M.; Bagdasarian, M. M.; Coleman, S.; and Timmis, K. N. 1979. New vector plasmids for gene cloning in *Pseudomonas*. In *Plasmids of medical, en-*

162   Genetic Engineering

vironmental and commercial importance, ed. K. N. Timmis and A. Pühler, pp. 411–22. Amsterdam: Elsevier/North-Holland Biomedical Press.

Beggs, J. D. 1978. Transformation of yeast by a replicating hybrid plasmid. *Nature* (London) 275:104–9.

Bolivar, F.; Rodriguez, R. L.; Greene, P. J.; Betlach, M. C.; Heyneker, H. L.; Boyer, H. W.; Crosa, J.; and Falkow, S. 1977. Construction and characterization of new cloning vehicles. II. A multipurpose cloning system. *Gene* 2:95–113.

Broach, J. R.; Strathern, J. N.; and Hicks, J. B. 1979. Transformation in yeast: development of a hybrid cloning vector and isolation of the CANI gene. *Gene* 8: 121–33.

Chakrabarty, A. M.; Mylroie, J. R.; Friello, D. A.; and Vacca, J. G. 1975. Transformation of *Pseudomonas putida* and *Escherichia coli* with plasmid-linked drug-resistance factor DNA. *Proc. Nat. Acad. Sci. (USA)* 72:3647–51.

Cohen, S. N.; Chang, A. C. Y.; Boyer, H. W.; and Helling, R. B. 1973. Construction of biologically functional bacterial plasmids in vitro. *Proc. Nat. Acad. Sci. (USA)* 70:3240–44.

Gantotti, B. V.; Patil, S. S.; and Mandel, M. 1979. Genetic transformation of *Pseudomonas phaseolicola* by R-factor DNA. *Molec. Gen. Genet.* 174:101–3.

Gryczan, T.; Shivakumar, A. G.; and Dubnau, D. 1980. Characterization of chimeric plasmid cloning vehicles in *Bacillus subtilis*. *J. Bacteriol.* 141:246–53.

Holloway, B. W. 1965. Variations in restriction and modification following increase of growth temperature of *Pseudomonas aeruginosa*. *Virology* 25:634–42.

Lethbak, A.; Christiansen, C.; and Stenderup, A. 1970. Bacterial genome sizes determined by DNA renaturation studies. *J. Gen. Microbiol.* 64:377–80.

Manson, M. D., and Yanofsky, C. 1976. Tryptophan operon regulation in interspecific hybrids of enteric bacteria. *J. Bacteriol.* 126:679–89.

Nagahari, K. 1978. Deletion plasmids from transformants of *Pseudomonas aeruginosa trp* cells with the RSF1010-*trp* hybrid plasmid and high levels of enzyme activity from the gene of the plasmid. *J. Bacteriol.* 136:312–17.

Nagahari, K.; Koshikawa, T.; and Sakaguchi, K. 1979. Expression of *Escherichia coli* tryptophan operon in *Rhizobium leguminosarum*. *Molec. Gen. Genet.* 171:115–19.

Nagahari, K., and Sakaguchi, K. 1978. RSF1010 plasmid as a potentially useful vector in *Pseudomonas* species. *J. Bacteriol.* 133:1527–29.

Nagahari, K.; Sano, Y.; and Sakaguchi, K. 1977. Derepression of *E. coli trp* operon on interfamilial transfer. *Nature* (London) 266:745–46.

Nagahari, K.; Tanaka, T.; Hishinuma, F.; Kuroda, M.; and Sakaguchi, K. 1977. Control of tryptophan synthetase amplified by varying the numbers of composite plasmids in *Escherichia coli* cells. *Gene* 1:141–52.

Sano, Y., and Kageyama, M. 1977. Transformation of *Pseudomonas aeruginosa* by plasmid DNA. *J. Gen. Appl. Microbiol.* 23:183–86.

Tanaka, T., and Sakaguchi, K. 1978. Construction of a recombinant plasmid composed of *B. subtilis* leucine genes and a *B. sublitis* (natto) plasmid. Its use as cloning vehicle in *B. subtilis* 168. *Molec. Gen. Genet.* 165:269–76.

Vogel, H. J., and Bonner, D. M. 1956. Acetylornithinase of *Escherichia coli*: partial purification and some properties. *J. Biol. Chem.* 218:97–106.

12

EMERGING TOOLS FOR IN VITRO AND IN VIVO GENETIC MANIPULATIONS IN PHYTOPATHOGENIC PSEUDOMONADS AND OTHER NONENTERIC GRAM-NEGATIVE BACTERIA

Nickolas J. Panopoulos

INTRODUCTION

Phytopathogenic pseudomonads are a large and heterogeneous group of bacteria causing economically important diseases in a wide variety of crop plants. Members of this group are often highly specialized pathogens infecting only one host plant species or single varieties or cultivars within, although some have a rather broad host range. The diseases themselves cover almost the entire range of symptomatological types encountered with bacterial plant pathogens: leaf spots, blights, vascular wilts, soft rots, knots, cankers, among others.

A number of these organisms, notably *Pseudomonas syringae* pathovars, are also noted for their ability to produce phytotoxins, several of which have been studied chemically as well as in relation to their role in disease (Durbin, forthcoming). *P. syringae* pathovars also have the peculiar ability to catalyze ice nucleation in supercooled water, a property that has been linked to frost damage on agricultural crops at relatively warm subzero temperatures (Lindow 1977). One member, *Pseudomonas syringae* pv. *savastanoi*, produces the plant growth hormone indoleacetic acid (Kosuge,

The active participation of Diane Sandlin, Brian Staskawicz, Mamoru Sato, Sasha Peters, Mary Honma, and Cindy Orser is acknowledged. Thanks are extended to A. Stepanov, A. M. Chakrabarty, R. Meyer, and G. Ditta for providing bacterial strains and plasmids. Portions of this work were supported by Grant no. 5901-0410-0241-0 from the USDA Competitive Grants Program.

Haeskett, and Wilson 1966). Thus, the group as a whole offers a wide range of possibilities for basic studies of various traits of phytopathological interest, such as race-specific pathogenicity, physiology of different types of disease symptoms, ice nucleation, toxin biosynthesis and role in disease, immunity of the pathogens to their own toxins, and evolution and potential for transmission of pathogenic traits among members of the group having varying degrees of genetic relatedness.

Members of the leaf-spotting group offer additional advantages that may be usefully exploited in basic studies of pathogen-host interactions. The intercellular spaces of leaves where the bacteria normally multiply are readily accessible to uniform inoculation by vacuum infiltration with bacterial suspensions at predetermined inoculum density. Bacterial cells administered in this way can be shown to give rise to leaf lesions with a probability approaching 1.0 (Panopoulos and Schroth 1974). Accurate quantitation of infection parameters and construction of detailed pathogen growth curves in planta can thus be obtained. The localized nature of the leaf lesions makes possible the simultaneous observation of many thousands of individual infection events in a small number of plants. This permits the isolation of extended-host-range or hypervirulent mutants (Panopoulos and Staskawicz, unpubl.), and perhaps genetic recombinants, from a large population of cells by using the leaf as a selective medium.

A main drawback in basic studies into the pathogenic and other traits of these bacteria has been the lack of understanding of their molecular genetics. Serious efforts to develop tools for genetic analysis in some members of the group began in the early 1970s (for review, see Lacy and Leary [1979]). (In *Pseudomonas aeruginosa*, which is considered by some investigators to be a quasi pathogen in plants, conjugational and transductional systems have been available for some time [Holloway, Krishnapilli, and Morgan 1979] and, more recently, methods for transformation and plasmid vectors suitable for deoxyribonucleic acid (DNA) cloning have been developed [Bagdasarian et al. 1979; Mercer and Loutit 1979; Nagahari and Sakaguchi 1978; see also Nagahari, Chapter 11]. However, the phytopathogenic capabilities of the strains used in molecular genetic studies are unknown, as is the applicability of the various genetic tools above to strains that are claimed to be pathogenic on plants.)

As a result of continuing efforts by several investigators, significant progress on the genetics of "bona fide" phytopathogenic species has been made in several areas. For example, a substantial body of knowledge has accumulated on the ability of different members of the group to exchange plasmids by conjugation (Lacy and Leary 1979; Panopoulos et al. 1978). The indigenous plasmids of several members have been described and characterized (Curiale and Mills 1977; Gantotti, Patil, and Mandel 1979a; Gonzalez and Vidaver 1979a,1979b,1979c; Panopoulos, Staskawicz, and

Sandlin 1979; Sands et al. 1979). In at least one instance *(P. s. savastanoi)*, the involvement of a plasmid in virulence and production of a pathologically significant metabolite (indoleacetic acid) has been cleary demonstrated (Comai and Kosuge 1980) (similar suggestions have been made for several phytotoxins, but the data are not conclusive; see Panopoulos and Staskawicz [forthcoming] and Gonzalez C. F. [1980: personal communication]). A partial genetic map of one member of the group has been constructed (Lacy and Leary 1979), and the transformability of several members has been reported (Gantotti et al. 1979b; Gross and Vidaver 1980; Lê, Leccas, and Boucher 1979; Panopoulos and Staskawicz, forthcoming; and this chapter). Plasmid replicons suitable for the cloning and propagation of recombinant DNA molecules in these organisms have been identified (Panopoulos 1979; Panopoulos et al. 1978; Panopoulos, Staskawicz, and Sandlin 1979; and this chapter).

Further progress in molecular genetic analysis of phytopathologically significant properties in those bacteria would benefit greatly from a concentration of efforts to selected model pathogens. However, it is unlikely that the diverse phenomena associated with different diseases can be adequately studied in any single species or outside the genetic background of the pathogens involved by utilizing, for example, DNA cloning techniques. It is for this reason that a broad range of genetic tools applicable to several different species is desirable.

This chapter presents recent progress toward the development and application of new molecular genetic techniques to phytopathogenic *Pseudomonas*. The subjects covered include: (1) plasmid replicons as cloning vectors useful for the development of DNA cloning systems in these bacteria, (2) construction of improved derivatives of plasmid RSF1010 suitable for cloning of DNA fragments generated by a variety of restriction endonucleases, and (3) use of specially constructed hybrid plasmids as suicidal carriers suitable for the introduction of transposable genetic elements (Tn) and for the construction of Hfr-type strains in these bacteria. The above tools are potentially useful for these purposes in many gram-negative bacteria outside the phytopathogenic *Pseudomonas* group. Detailed accounts of the results discussed here are published elsewhere (Panopoulos et al., in press; Sandlin and Panopoulos, in preparation; Sato, Staskawicz, and Panopoulos, forthcoming).

DNA CLONING SYSTEMS

DNA cloning can be exceedingly useful in the genetic analysis of phytopathogenic traits. The isolation of genes on small replicons greatly facilitates detailed structural and functional analysis, identification of transcriptional and translational products, functional complementation tests, and selective mutagenesis and makes DNA sequencing possible. The task of isolating a particular gene from a prokaryotic source is relatively simple when the gene resides on a small replicon or specifies a selectable phenotype that is expressed in the cloning host or when a specific biochemical, molecular, or immunological probe is available (see Strathern, Chapter 13). For some pathologically significant traits these requirements are easily met. For instance, the involvement of plasmids in pathogenicity has been demonstrated for certain plant pathogenic bacteria: *Agrobacterium tumefaciens* (see Chapter 8), *A. rhizogenes* (Moore, Warren, and Strobel 1979), and *Pseudomonas syringae* pv. *savastanoi* (Comai and Kosuge 1980).

However, compelling arguments that small replicons are likely to be the exclusive or frequent carriers of genes determining pathogenicity and related traits, in these or other species, cannot be made. Furthermore, for many of these traits specific biochemical or other tests are neither available nor obvious. In these cases, cloned DNA fragments carrying the genes in question can be unambiguously identified only through genetic complementation in the appropriate genetic background, namely, virulent or host-range mutants of the pathogenic strain in question. Since virulence, pathogenicity, and host range are most likely controlled by a constellation of genes that must function in concert for the full expression of these traits, it is unlikely that *Escherichia coli*, the principal prokaryotic cloning host in current use, will satisfy the entire range of experimental needs envisioned in the use of recombinant DNA technology to study phytopathogenicity. The capability of introducing recombinant molecules in phytopathogenic species is, therefore, considered essential for basic research purposes in plant pathology.

More practical reasons for the development of host-vector systems in phytopathogenic prokaryotes can also be stated. The use of plant pathogens as biological weed control agents (Templeton, TeBeest, and Smith 1979) and avirulent (or hypovirulent) strains of some phytopathogens as biological antagonists (Day and Dodds 1979; S. E., Lindow 1980: personal communication) are already being practiced, although on a restricted scale at present. Conceivably, weak pathogens or antagonists can be made more effective by genetic manipulation. As in the case of insect biocontrol agents (see Chapters 14 and 15), the deployment of pathogens that have been genetically engineered by in vitro methods in the field, now prohibited under the National Institutes of Health (NIH) Guidelines, may be waived if biosafety requirements can be satisfied. Finally, some phytopathogenic bacteria produce commercially useful products — for example, xanthan, a complex polymer of *Xanthomonas*. Genetic amplication of the genes encoding rate-limiting enzymes of the biosynthetic pathway by cloning on multicopy plasmid vectors can theoretically lead to increased production of the end product in question.

Our efforts to develop DNA cloning systems in plant pathogenic *Pseudomonas* have proceeded along the following lines: identification of vector replicons suitable for the propagation of recombinant molecules in these organisms, construction of improved derivatives of these vectors to broaden the range of restriction endonuclease sites available for DNA fragment insertion and to permit identification of recombinant clones by insertional inactivation, and assessment of the feasibility of introducing the recombinant molecules in various species by transformation or conjugative mobilization. Most of these experiments have been conducted with selected members of the group but are likely to be more generally applicable to other members as well as to other nonphytopathogenic gram-negative bacteria.

Vector Replicons

Among the plasmids that appeared promising as potential cloning vectors for phytopathogenic pseudomonads were the broad-host-range plasmids of the *incP*-1 group, namely, RP1, RP4, RK2 (Boistard and Boucher 1979; Panopoulos et al. 1978), which appear to be identical to each other by molecular hybridization and restriction endonuclease cleavage criteria (Burkardt, Reiss, and Pühler 1979) and plasmid RSF1010. Derivatives of the ColE1, ColE1-like, or P15A plasmid replicons, which include some of the most advanced plasmid vectors (Bolivar et al. 1977; Chang and Cohen 1978), do not appear capable of stable autonomous replication in nonenteric bacteria (Bagdasarian et al. 1979; Ditta et al. 1980; Ditta, Chapter 10; unpubl. data from this laboratory). One report suggesting that pBR322 can replicate in *P. s. phaseolicola* (Gantotti, Patil, and Mandel 1979b) has not been substantiated (this laboratory, unpubl.). The molecular vehicle properties of *incP*-1 plasmids (Barth 1979a;1979b; Hedges, Cresswell, and Jacob 1976; Helinski 1977; Jacob, Cresswell, and Hedges 1976; Jacob and Grinter 1975; Meyer, Figurski, and Helsinki 1975,1977) and their reduced-size derivatives (Ditta et al. 1980; Khan et al. 1979; Ditta, Chapter 10) and of RSF1010 (Bagdasarian et al. 1979; Nagahari and Sakaguchi 1978; Nagahari, Chapter 11) have been established.

The conjugational proficiency on *incP*-1 plasmids permitted the determination of their ability to replicate in phytopathogenic pseudomonads, as in most other gram-negative bacteria, early in the history of genetic studies in these species and before their transformability with purified plasmid DNA was demonstrated. The host range of these plasmids includes all members of the phytopathogenic *Pseudomonas* group studied to date, in addition to other plant pathogens (Lacy and Leary 1979; Panopoulos et al. 1978). The large size and conjugative properties of these plasmids have generally discouraged their use as vectors with plant pathogenic hosts. Smaller, conjugation-defective derivatives of RK2 (pRK248, pRK229) constructed by in vitro methods have been available for some time (Helinski 1977). However, their introduction and routine use as vectors in plant pathogenic bacteria was prohibited in the United States prior to January 1979 by the NIH Guidelines. For this reason alternative vector replicons for phytopathogenic *Pseudomonas* were sought.

RSF1010 is a small (5.5 Md), nonconjugative, naturally occurring plasmid specifying resistance to streptomycin (Sm^r) and sulfonamides (Su^r) and has useful molecular vehicle properties (Bagdasarian et al. 1979; Barth 1979b; Barth and Grinter 1974; Grinter and Barth 1976; Guerry, Embden, and Falkow 1974; Nagahari and Sakaghuchi 1978). Nagahari and Sakaguchi (1978) first showed that this plasmid was capable of stable autonomous replication in *Pseudomonas aeruginosa* and *P. putida*. These findings prompted us, in 1978, to investigate whether RSF1010 was capable of autonomous replication in phytopathogenic pseudomonads. In the absence, at that time, of a suitable method to render these bacteria competent for transformation, we initially sought to determine whether RSF1010 could be "mobilized" in *trans* (that is, without prior recombination) by the conjugation-proficient plasmid RK2 (Ap^r, Km^r, Tc^r), which was known to be capable of mating with phytopathogenic pseudomonad recipients (Panopoulos et al. 1978).

It was in this manner that the plasmid was first shown to be capable of stable replication in pv. *phaseolicola* and other members of the *P. syringae* group. In matings between *E. coli* C600 (RSF1010, RK2) and various *Pseudomonas* recipients, transconjugants were obtained that contained RSF1010 but not RK2. The relative percentage of such transconjugants compared with those that acquired both RSF1010 and RK2 in these crosses varied both with the time allowed for mating and with the relative frequency with which RK2 was transferred to them from *E. coli* (RK2) donors. Short conjugation time and low recipient ability for RK2 readily permitted the isolation of transconjugants that had acquired RSF1010 but not RK2 markers. The presence of RSF1010 and absence of RK2 in the transconjugants, confirmed by agarose gel electrophoresis (Panopoulos, Staskawicz, and Sandlin 1979), demonstrated the autonomous replicative ability of RSF1010 in several *P. syringae* pathovars. Those and subsequent experiments involving either RK2 or pBPW1, a naturally occuring conjugative plasmid from *P. s. tabaci* (Staskawicz, Sato, and Panopoulos 1981), as mobilizing plasmids, as well as transformation with purified plasmid DNA, have extended the host range of RSF1010 to all phytopathogenic pseudomonads thus far examined (Table 12.1).

TABLE 12.1 ————————————————
Host Range of RSF1010 and Derivatives
Among Phytopathogenic *Pseudomonas*

Bacteria	Number of Strains Examined	Method of Introduction
RSF1010		
Pseudomonas syringae pathovars		
pv. *angulata*	1	M, pBPW1
pv. *glycinea*	1	T[a]
pv. *mori*	12	M, pBPW1
pv. *phaseolicola*	3	T, M, RK2
pv. *savastanoi*	2	T
pv. *syringae*	1	M, pBPW1, T
pv. *tabaci*	1	T
pBP-S1[b]		
pv. *phaseolicola*	1	T
pBP-S2		
pv. *phaseolicola*	1	T
pBP-S3		
pv. *phaseolicola*	1	M, pAS8Tc ^{s}*rep-1*::Tn7
pv. *syringae*	1	T[a]

[a]Orser (unpubl.).

[b]D. Foulds has also introduced this plasmid in *Myxococcus xanthus* by transformation (1980: personal communication).

Notes: M = mobilization by plasmid indicated; T = transformation.

Source: Compiled by the author.

RSF1010 is stably inherited in pv. *phaseolicola* and other *P. syringae* pathovars through several sequential single-colony purifications in the absence of selection and during strain storage. Thus, the ability of RSF1010 to be maintained stably as an autonomous replicon in phytopathogenic pseudomonads makes it potentially useful as a cloning vector for the propagation of in vitro constructed recombinant molecules in these bacteria.

Transformability of Phytopathogenic *Pseudomonas* Strains

A desirable, although not essential, prerequisite for the establishment of a cloning system in a new host is efficient and reproducible transformation. Several investigators had reported in preliminary form apparent genetic transformation (restoration to prototrophy) of different phytopathogenic pseudomonads utilizing total homologous, presumably linear DNA (Coplin, Sequeira, and Hanson 1974; Twiddy and Liu 1972; Vasile'va and D'Yakov 1970), but no further reports by these authors have appeared on the subject. In more recent work, Boucher and Sequeira (1978) and Lê, Leccas, and Boucher (1979) reported deoxyribonuclease-sensitive (DNase-sensitive) apparent cotransformation for unselected prototrophic markers in *P. solanacearum* with total crude, homologous DNA preparation.

Our demonstration of the autonomous replicative ability of RSF1010 in *P. s. phaseolicola (vide supra)* permitted the first unequivocal demonstration of genetic transformation in this bacterium (Gantotti, Patil, and Mandel 1979b) (see Table 12.2). We have subsequently used this plasmid as a probe molecule to determine the transformability of several *P. syringae* pathovars, to compare the effectiveness of competence-inducing treatments and other variables for maximizing transformant recovery, and to search for strains in which the frequency of transformation is sufficiently high to tolerate the substantial decrease in the transforming efficiency of plasmid vectors experienced in routine cloning work. RSF1010 DNA was isolated from *P. putida* AC808 (RSF1010) by two cycles of dye-buoyant density centrifugation followed by extensive dialysis against Tris-EDTA buffer (10 mM Tris-hydroxylaminomethane, 1 mM sodium ethylenediaminotetraacetate, pH 7.5). In the presence of $CaCl_2$ (Chakrabarty et al. 1975), transformant yields were in excess of 10^3 per μg DNA with pv. *phaseolicola* strain NPPH3006 and 10^5 per μg with NPPH4001 after 2 hours of growth in Sm-free medium. With longer incubation in Sm-free medium, the number of transformants increased considerably, partly because of multiplication. The efficiency of transformation was further increased to more than 10^6 trans-

TABLE 12.2————————————————————
Transformation Frequencies for Two Strains of pv. *phaseolicola* After Treatment with $CaCl_2$ and $MgCl_2$

Strain	Time in Sm-free Medium (hr)	Transformants/μg DNA	
		$CaCl_2$ (100 mM)	$MgCl_2$ (250 mM)
NPPH4001	0.5	1.2×10^3	
	2	1.9×10^4	1.9×10^6
	4	4.3×10^5	
	24	8.0×10^6	
NPPH3006	0.5	1.6×10^2	
	2	1.1×10^3	1.9×10^3
	4	1.6×10^3	
	24	2.4×10^4	

Notes: RSF1010 was isolated from *P. putida* AC808(RFS1010). Transformants were plated on minimal glucose salts medium supplemented with 20 mg/ml streptomycin.
Source: Compiled by the author.

formants per μg DNA in NPPH4001 by substituting $MgCl_2$ for $CaCl_2$, as reported for other *Pseudomonas* (Mercer and Loutit 1979). The increase may be due in part to the avoidance of extensive cell coagulation frequently observed with this and, to a lesser extent, with other strains upon addition of $CaCl_2$. The effectiveness of $MgCl_2$ in other strains has not been determined. NPPH4001 may be a more efficient acceptor of foreign DNA, as it has been selected as a hyperfertile conjugational recipient for plasmid RP1 following chemical mutagenesis of strain HB10Y (Mindich, Cohen, and Weisburd 1976). However, the basis for its increased recipient ability in conjugation has not been determined and a possible relationship between this property, and the higher transformability reported here is only speculative.

Sm^r colonies in the control cultures (no DNA, $MgCl_2$, or $CaCl_2$) were very infrequent (10^{-8} to 10^{-9} per cell plated), indicating that the process was both DNA and divalent cation dependent. Sm^r transformants were also resistant to sulfonamides and contained a new plasmid species similar to RSF1010 in its electrophoretic mobility in agarose gels. These properties remained stable after at least 10 single-colony purifications. The results, therefore, show that $CaCl_2$ or $MgCl_2$ treatment makes pv. *phaseolicola* competent for transformation at high frequency.

The transformation frequencies obtained with strain NPPH4001 in our experiments are higher than those reported for RSF1010 in *P. aeruginosa* and *P. putida* by Bagdasarian et al. (1979) and comparable to those obtained with RSF1010 in an *E. coli* $r_k^- m_k^+$ strain by these authors and by Gross and Vidaver (1981) in *P. syringae*. These authors used a procedure involving $CaCl_2$, glycerol, and heat shock at 45°C. With strain NPPH3006 the transformation frequency, although lower than with NPPH4001, was considerably higher than that reported in another study with strain G50 (the parent strain of NPPH3006) and other strains of pv. *phaseolicola* using RSF1010 DNA from *P. putida* (Gantotti, Patil, and Mandel 1979b).

Construction of RSF1010 Derivatives

RSF1010 has single cleavage sites for the *Eco*RI, *Bst*EII, *Hpa*I, and *Pvu*II restriction endonucleases and two *Pst*I sites that are separated by 0.46 Md (Bagdasarian et al. 1979; Barth 1979b; Barth and Grinter 1974; Grinter and Barth 1976). Both the *Eco*RI and *Pst*I sites have been used for insertion of DNA fragments (Bagdasarian et al. 1979; Nagahari et al. 1977). The latter, at least one of which is located within the Su^r gene (Bagdasarian et al. 1979), permits identification of recombinant molecules by insertional inactivation. There are no sites for *Bam*HI, *Hind*III, *Kpn*I, *Bgl*II, *Sma*I/*Xma*I, *Xba*I (Bagdasarian et al. 1979), or *Xho*I (Sandlin and Panopoulos, in preparation) on RSF1010.

These attributes have been exploited to construct several derivatives of RSF1010 that improve its suitability as a cloning vector in several ways. One derivative, pBP-S1, consists of RSF1010 and a 4.5 Md fragment from plasmid pML2 (Hershfield et al. 1974) specifying kanamycin resistance fused at their *Eco*RI sites. The inserted fragment contains single *Hind*III, *Sal*I, and *Xho*I sites, which are absent from RSF1010 (Fig. 12.1). pBP-S2 and pBP-S3 are two other derivatives constructed by fusing RSF1010 with

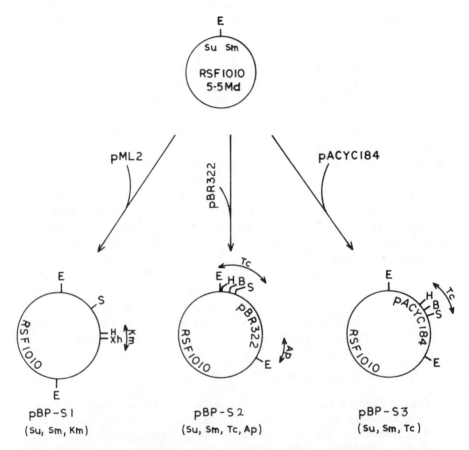

FIG. 12.1. Schematic drawing of the pedigree of three RSF1010 derivatives. All constructions involved the fusion of *Eco*RI-cleaved molecules. The location of single restriction endonuclease sites, as well as the two *Eco*RI sites, and the markers of the inserted fragments are shown in each derivative. The orientation of the inserts relative to RSF1010 coordinates has not been determined. Abbreviations: **E, B, H, S, Xh**: *Eco*RI, *Bam*HI, *Hind*III, *Sal*I, and *Xho*I, respectively; **Ap, Sm, Su, Tc**: ampicillin, streptomycin, sulfonamide, and tetracycline resistance, respectively.

plasmids pBR322 (Bolivar et al. 1977) and pACYC184 (Chang and Cohen 1978), respectively, at their single EcoRI sites. These derivatives contain single cleavage sites for HindIII, BamHI, and SalI restriction endonucleases (Fig. 12.1). pBP-S1 and pBP-S3 still permit cloning into the PstI sites carried by RSF1010. The combination of three markers on these vectors should permit the selection of transformants by either Sm or Tc resistance, elimination of mutants by testing for simultaneous resistance to sulfonamides, and the identification of recombinant clones by insertional inactivation: insertions into the PstI site(s) of pBP-S1 and pBP-S3, the HindIII and XhoI sites of pBP-S1, and the HindIII, BamHI, and SalI sites of pBP-S2 and pBP-S3 should inactivate Su^r, Km^r, and Tc^r, respectively. We have used the pBP-S3 vector successfully to clone HindIII-generated fragments of pv. phaseolicola total DNA (Sandlin and Panopoulos, in preparation). A sample of these recombinant plasmids, expressing a Tc^{s}Sm^r phenotype, was examined by gel electrophoresis and found to contain inserts of up to 18 kb in size (Fig. 12.2).

The three vectors above provide usefull alternatives to the RK2-derived vectors (Ditta et al. 1980; Khan et al. 1979; Ditta, Chapter 10) with which they should be compatible but from which they presumably differ in copy number. A reduction in size, currently being pursued in this laboratory, should further improve their suitability. Vector pBP-S1 provides for direct selection of recombinant plasmids if used with a streptomycin/paromomycin-dependent host.

An important consideration for the use of RSF1010 derivatives as cloning vectors is their ability to be maintained in the cloning host. Although Nagahari et al. (1977) did not attempt to introduce RSF1010 hybrid plasmids in Pseudomonas, Bagdasarian et al. (1979) were able to introduce all but one of their derivatives in P. aeruginosa and P. putida. The exception was pKT210, the RSF1010-pBR322 hybrid constructed by PstI fusion. In our studies we have successfully introduced pBP-S1, pBP-S2, and pBP-S3 by transformation or mobilization in various phytopathogenic pseudomonads. The plasmids are stably maintained in the presence of antibiotics.

Biosafety Considerations

In contemplating the use of phytopathogenic pseudomonads as recipients for in vitro constructed recombinant molecules, it is desirable to know whether the vector replicon is mobilized by indigenous fertility systems that may be present in the host cells. Indigenous plasmids greater than 20 Md in size and, therefore, potentially conjugative, have been detected in a number of P. syringae pathovars (Curiale and Mills 1977; Gantotti, Patil, and Mandel 1979a; Panopoulos, Staskawicz, and Sandlin 1979; Sato, Staska-

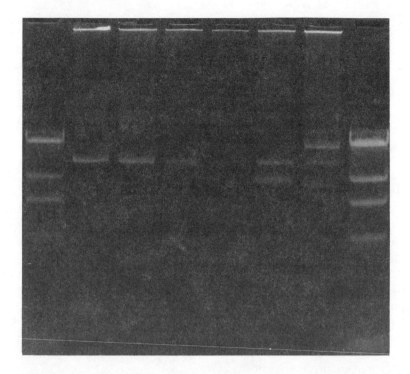

FIG. 12.2. Cloning of pv. *phaseolicola* DNA in the *Hind*III site of vector pBP-S3. Inner lanes contain representative recombinant plasmids from Sm^{r}Tc^s clones; outer lanes contain phage λ DNA fragments. All samples were digested with *Hind*III prior to electrophoresis. The DNA source strain was Tox$^-$. The cloning host was *E. coli* SK1592. The uppermost band in the inner lanes (except the last one on the right) is the vector. The last lane on the right apparently contains an insert with internal *Hind*III sites. Electrophoresis was performed in Tris-borate buffer, with a 0.7 percent horizontal agarose gel. Staining was with ethidium bromide. Smaller bands are diffuse because the photograph was taken one day after electrophoresis. Cloning was performend by D. E. Sandlin.

wicz, and Panopoulos, forthcoming; Staskawicz, Sato, and Panopoulos 1981).

Mobilization of RSF1010 from pv. *phaseolicola* NPPH4001 and NPPH3006, pv. *tabaci* BR2, pv. *syringae* 31, and several pv. *mori* strains was tested by overnight matings on membrane filters with the restriction-deficient *E. coli* strain SK1592. Only pv. *tabaci* was able to donate RSF1010 readily to *E. coli* in these crosses (frequency was approximately 10^{-1} per donor). Mobilization from other strains was either very low (10^{-6} to 10^{-9} per cell plated in pv. *mori* strains) or undetected less than 10^{-8} per cell plated with pv. *phaseolicola* NPPH4001 or NPPH3006). Thus, with the exception of pv. *tabaci* none of the other strains appear to have fertility systems capable of efficiently mobilizing RSF1010. The fertility system of pv. *tabaci* BR2 is attributed to the pBPW1 plasmid present in this strain (Sato, Staskawicz, and Panopoulos, forthcoming).

IN VIVO GENETIC MANIPULATIONS

As indicated earlier, capability for in vivo genetic studies in phytopathogenic pseudomonads has been somewhat restricted in the past. The demonstration of transformation for an increasing number of strains (Comai and Kosuge 1980; Gantotti, Patil, and Mandel 1979b; Gross and Vidaver 1981; Lê, Leccas, and Boucher 1979; and this chapter) greatly increases the prospect of detailed genetic analysis. One area of concentrated activity at present concerns the possible association between plasmids and toxigenicity or other virulence properties. Recent studies in this laboratory have been directed toward the development of efficient tools for transposon mutagenesis, which would facilitate these studies, and for the construction of Hfr-type strains useful in chromosome mapping.

Direct Selection for Transposon Insertions

Transposable drug-resistance elements (transposons or Tn) are useful for a variety of genetic manipulations in bacteria: mutagenesis, phenotypic marking of plasmids, mapping of genes with unselectable phenotype, and construction of conjugationally "fertile" donor strains, among others (Kleckner, Roth, and Botstein 1977). Tn elements are generally introduced into a new bacterial strain as a part of a carrier replicon, usually plasmid or phage. Transposition into other resident genomic elements subsequently occurs with frequencies that depend on the Tn element itself, the host strain, and, often, on culture conditions (for example, temperature), and the practical problem is to identify clones in which such events have occurred. This task is greatly simplified if the carrier replicon is unable to establish itself in the recipient strain or can be easily eliminated after its introduction. In such cases, clones that retain the transposon are obtained by direct selection with an appropriate antibiotic. Several types of such carriers are now available: plasmids with thermosensitive replication, such as F'_{t_s} plasmids (Kleckner, Roth, and Botstein 1977) and plasmid pSC201 (Kretschmer and Cohen 1977); phage λ and P22 derivatives that do not lysogenize and/or that do not kill the recipient cells (Kleckner, Roth, and Botstein 1977); and the "suicidal" derivatives of *incP*-1 plasmids, such as RP4 and pPH1JI, carrying bacteriophage Mu (Beringer et al. 1978; Van Vliet et al. 1978). With the exception of the latter two carriers above, the others have been used exclusively in enteric bacteria. There are no reports that these carriers are suitable for direct selection of Tn insertions in phytopathogenic *Pseudomonas*.

Use of replicons of enteric origin as Tn carriers for nonenteric bacteria, including phytopathogenic *Pseudomonas*, may be impractical for various reasons. Transmission barriers are likely, especially for the phages. Transformation, a means for overcoming these barriers, occurs in relatively few species and, with few exceptions, is a low-frequency event. Many species do not grow at the temperatures necessary for curing of the thermosensitive carriers available (40–42°C). This limitation also applies to the thermosensitive broad-host-range plasmids, such as pEG1 (that is, $RP4_{ts}$; Danilevich et al. 1978), in these bacteria.

Carriers consisting of a broad-host-range plasmid replicon with phage Mu inserted in their genome are attractive for Tn insertion purposes in these bacteria, although they have been used in relatively few species thus far. These plasmids have greatly reduced transmission frequencies, by comparison with their non-Mu counterparts, in integeneric crosses (Boucher, Barate de Bertalmio, and Dénarié 1977; Dénarié et al. 1977; Van Vliet et al. 1978). The basis underlying their suicidal behavior appears to involve the action on DNA restriction barriers in heterologous crosses and formation of extensive Mu-promoted deletions on the plasmid genome. The function of at least one phage Mu gene *(r23)* (Boucher, Barate de Bertalmio, and Dénarié 1977; Dénarié et al. 1977; Van Vliet et al. 1978) appears necessary for suicide.

These principles are sufficiently general to suggest a broad use of Mu-containing *incP*-1 plasmid derivatives as Tn carriers in diverse bacteria. Indeed, RP4::Mu::Tn7 has proved useful in isolating Tn7 insertion mutants in *Agrobacterium tumefaciens* (Van Vliet et al. 1978). However, expression of the phage Mu genome evidently varies with the recipient strain, and, in one instance, involving the plant pathogen *Pseudomonas solanacearum*, RP4::Mu plasmids did not express their suicidal properties, even though the phage Mu genome was clearly expressed (that is, Mu particles were formed) in the recipient (Boucher, Barate de Bertalmio, and Dénarié 1977; Dénarié et al. 1977). Furthermore, RP4::Mu::Tn7 proved unsatisfactory for the isolation of Tn7 insertions in *Rhizobium leguminosarum* (Beringer et al. 1978), even though the plasmid expressed suicidal behavior in this host. Plasmid pJB4JI, a derivative of another *incP*-1 plasmid, pH1JI, containing Mu and Tn5 (Beringer et al. 1978), has been used successfully in several *Rhizobium* species to generate Tn5 insertions (Beringer et al. 1978; Long et al., Chapter 9 of this book). However, some of the mutants obtained in *R. meliloti* contained parts of the suicide plasmid, in addition to Tn5, which complicated genetic interpretation (F. Ansubel 1981: personal communication).

In our genetic investigations with phytopathogenic pseudomonads, we have used successfully a conditional host-dependent plasmid carrier that has narrow-host-range replication but broad-host-range conjugational properties to obtain Tn7 insertions in several members of this group. Because of its special construction and properties, this plasmid may be more generally useful as a Tn carrier for many bacteria, as well as for other transposons, in addition to those used in this study.

Plasmid pAS8Tc^{s}rep-1::Tn7, here referred to as pAS8Rep-1, is a tetra-cycline-sensitive derivative of plasmid pAS8 (that is, RP4-ColE1, fused at their EcoRI sites [Sakanyan et al. 1978]) carrying the Tn7 transposon inserted in a gene necessary for RP4-specific replication (A. Stepanov 1979: personal communication). The plasmid is conjugation proficient (Tra^+_{RP4}) and replicates via the ColE1 replication system, as shown by its dependence on a functional polA gene in E. coli (Sato, Staskawicz, and Panopoulos, submitted; A. Stepanov 1979: personal communication). ColE1 and ColE1-like plasmids are generally considered narow-host-range replicons as they cannot be experimentally introduced or stably maintained in bacteria outside the Enterobacteriaceae, such as Pseudomonas aeruginosa, Rhizobium spp. (Bagdasarian et al. 1979; Ditta, Chapter 10) or Pseudomonas syringae pv. phaseolicola (our unpublished results; a report [Gantotti, Patil, and Mandel 1979b] suggesting the contrary could not be substantiated). Accordingly, we reasoned that pAS8Rep-1 should not be able to replicate or maintain itself in phytopathogenic pseudomonads following conjugation but that transposable elements carried on it might be rescued by selection for resistance to the appropriate antibiotic.

We investigated the suitability of this plasmid for isolating Tn7 transposition mutants primarily because the linked resistance to two antibiotics, trimethoprim (Tp^r) and streptomycin (Sm^r), specified by this transposon (Barth et al. 1976), facilitated the elimination of spontaneous mutants during selection. In addition to Sm^r and Tp^r the plasmid carries the PP4 markers for resistance to kanamycin and neomycin (Km^r/Nm^r) that is not transposable (M. H. Richmond 1978: personal communication) and for ampicillin (Ap^r), which is on transposon Tnl (Campbell et al. 1979; Hedges and Jacob 1974). Thus, it was possible to determine whether the carrier replicon was maintained in the recipients and, thereby, to confirm its conditional host-dependent replicative ability. The experiments involved overnight matings between a multiauxotrophic E. coli donor and phytopathogenic pseudomonads recipients on agar media or on Millipore filters. Transconjugants were selected by plating on minimal media (with rifampicin-resistant recipients, complex medium containing rifampicin was used) supplemented with the appropriate antibiotic.

Several strains of pv. phaseolicola and pv. mori and one strain of pv. tabaci were studied (Sato, Staskawicz, and Panopoulos, submitted). Colonies resistant to kanamycin occurred with very low frequency, not substantially higher than the frequency of spontaneous mutation to kanamycin resistance. This presumably reflects the inability of the ColE1 replication system to function effectively in *Pseudomonas*. By contrast, colonies resistant to streptomycin or trimethoprim occurred at considerably greater frequencies — up to 10^4-fold — than either those of Km^r colonies or of spontaneous Sm^r or Tp^r mutants. With some strains, essentially all (98.5 to 100 percent) transconjugant colonies selected for Sm^r were also resistant to Tp (and vice versa) but sensitive to kanamycin and ampicillin and contained no new plasmid species detectable by agarose gel electrophoresis. These results suggested that the $Sm^r Tp^r$ phenotype was acquired as a result of transportation of Tn7 following transfer and abortive replication of the carrier genome. Direct evidence for Tn7 insertions has been obtained by (1) identification of clones carrying insertions on indigenous plasmids (Fig. 12.3); some of these insertions cause Tra⁻ mutations on the pBPW1 plasmid

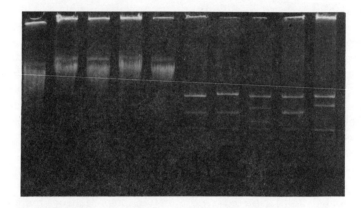

FIG. 12.3. Transposition of Tn7 to plasmid pBPW1 of *Pseudomonas tabacci* BR2 obtained through the use of the hose-dependent suicidal plasmid pAS8Tc^s *rep-1*::Tn7. DNAs in lanes 1 through 5 (from the left) are uncleaved; those in lanes 6 through 10 are digested with *Hind*III. Lanes 1 and 6 represent the wild type strain; all other lanes represent pBPW1::Tn7 insertion derivatives. The slower electrophoretic mobility of pBPW1::Tn7 plasmids (uppermost band in lanes 2, 3, 4, and 5) relative to pBPW1 (lane 1) and their altered *Hind*III fragment patterns (in the four rightmost lanes) relative to that of pBPW1 (lane 6 from the left) are consistent with Tn7 insertions in different locations. The two internal *Hind*III fragments of Tn7 are those absent in the cleared pBPW1 DNA but present in the other lanes. The smear in the uncleaved preparation is due to chromosomal DNA contamination, and the band just below the smear is an artifact experienced with the extraction method used. Cleavage of pBPW1 with *Hind*III produces seven fragments (the three smaller ones are faint in the picture). Electrophoresis conditions as in Fig. 12.2.

of pv. *tabaci* BR2 (Sato, Staskawicz, and Panopoulos, forthcoming); and (2) by recovering the Tn7 element from chromosomal locations after transposition onto various plasmids. In all cases the insertions were confirmed as due to Tn7 by restriction enzyme analysis of the target plasmids (Sato, Staskawicz, and Panopoulos, forthcoming; see also Fig. 12.4).

pAS8Rep-1 may be suitable as a general purpose Tn carrier for nonenteric gram-negative bacteria. In addition to Tn7 it carries transposon Tn1 (Ap^r). Ap^r mutants, presumed to be Tn1 insertions, have been obtained in pv. *phaseolicola* (Panopoulos, unpubl.). Presumably, derivatives carrying

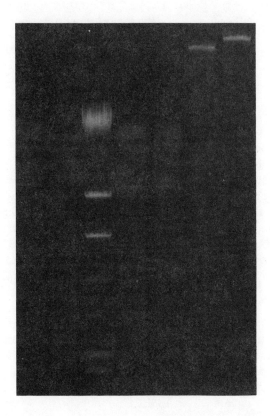

FIG. 12.4. Tn7 insertions in RP1-*lac* (plasmids designated pBP) obtained after conjugation between pv. *phaseolicola* [Chr::Tn7](RP1-*lac*) with *E. coli* MV12 (a *recA* strain). Samples, from right to left, are: pBP3, RP1-*lac* (both uncleaved), pBP3, RP1-*lac*, λ DNA, pBP6, pBP7, and pBP8 (all six cleaved with *Hind*III. Altered cleavage patterns are consistent with Tn7 insertions in various locations of RP1-*lac*. Some smaller fragments of the plasmids are not very evident in the photograph. Electrophoresis conditions as in Fig. 12.2.

other transposons could be constructed. Simultaneous (nonselected) transposition of Tn1 has not been observed among over 2,500 Tn7$^+$ transconjugants thus far examined, suggesting that presence of more than one transposon on this carrier should not complicate the genetic analysis of mutants generated by selection for one transposon only. Theoretically, the range of organisms in which pAS8Rep-1 may be useful covers any strain capable of participating in matings with RP4 or similar plasmid donors and does not support stable replication/maintenance of plasmid ColE1 or effectively complement the *rep-1* mutation, which renders the plasmid dependent on the ColE1 replication system. The first requirement is satisfied by most gram-negative bacteria. Efficient ColE1 replication appears restricted to enteric species. Neither *E. coli* nor the majority of strains within the phytopathogenic *P. syringae* pathovars thus far studied appear to provide efficient complementation for the *rep-1* mutation on pAS8Rep-1 (Sato, Staskawicz, and Panopoulos, forthcoming).

Direct Selection for Hrf-type Strains

The term *Hfr*, originally coined for strains carrying an F or F′ plasmid integrated into the bacterial chromosome, is expanded here to include strains carrying integrated genomes of transmissible R plasmids. Such strains, useful in chromosome mapping, have not been described in phytopathogenic *Pseudomonas*, although a naturally occurring plasmid of unknown fertility in pv. *phaseolicola* has been observed to undergo integration into the chromosome (D. Mills 1979: personal communication).

The properties of pAS8Rep-1 (*vide supra*) make it a suitable candidate for the construction of Hfr strains in phytopathogenic pseudomonads by direct selection. The nontransposable nature of Km^r provides a strong selection for plasmid maintenance, and integration may be facilitated by prior insertion of Tn7 or Tn1 into the chromosome, a task for which the plasmid is well suited. Furthermore, the inability of the ColE1 replication system to function in these bacteria reduces, in theory at least, the likelihood that simultaneous presence of pAS8Rep-1 as a free plasmid in these Hfr strains might be required for viability (Sasakawa and Yoshikawa 1980). Selection for Km^r transconjugants with pAS8Rep-1 donors occurs with very low frequency in phytopathogenic *Pseudomonas*. However, stable Km^r Ap^r Sm^r Tp^r transconjugants have been obtained in pv. *phaseolicola* and pv. *mori* carrying chromosomal Tn7 insertions (Sato and Panopoulos, unpubl.). Although analysis of these transconjugants is still in progress, at least some appear to contain no new plasmid species and are presumed to have chromosomally integrated pAS8Rep-1 sequences.

CONCLUSION

Several emerging tools for genetic studies in phytopathogenic pseudomonads were described. Although the data are still preliminary in many instances, it appears that genetic analysis in this group of bacteria can now proceed at an accelerated pace. The tools developed should also prove useful in other bacteria. RSF1010 is a broad-host-range replicon among phytopathogenic pseudomonads and has already been used to clone tryptophan monooxygenase, an enzyme involved in indoleacetic acid biosynthesis in *P. s. savastanoi* and to reintroduce the recombinant molecules in this organism (Comai and Kosuge 1980; L. Comai 1981: personal communication). pBP-S1 has been introduced into *Myxococcus xanthus*, a nonplant pathogen (D. Fould 1980; personal communication); pBP-S3 has been successfully used in this laboratory to establish a colony bank for pv. *phaseolicola*. These plasmids may be used either for direct cloning, when transformation frequencies are high, or as vector components in binary cloning systems similar to that described by Ditta et al. (1980) and in Chapter 10 of this book. pAS8Rep-1 or pRK2013 may be employed as mobilizing plasmids. Both have the broad mating properties of the *incP-1* fertility system, which mediates conjugation with most gram-negative bacteria, including phytopathogenic *Pseudomonas*, and mobilizes the RSF1010 replicon in trans. Their narrow replicative host range adds a useful biosafety feature. Mobilization of pBP-S3 by pAS8Rep-1 from *E. coli* to pv. *phaseolicola* in triparental matings occurs with a frequency of 10^{-3} per recipient (this laboratory, unpubl.), and the transconjugants do not maintain the mobilizing plasmid.

Our studies have demonstrated the suitability of pAS8Rep-1 for transposon mutagenesis in *Pseudomonas*. We (Orser and Panopoulos, unpubl.) as well as others (C. A. Carlson and J. L. Ingraham 1981: personal communication) have recently isolated Tn7-insertion auxotrophic mutants in *P. s. syringae* and *P. stutzeri*, respectively, using this carrier plasmid. Our preliminary data suggest that the plasmid may prove useful as a dual-purpose plasmid, permitting, in addition, the construction of Hfr-type strains useful in chromosome mapping. Another possible use for pAS8Rep-1, so far unexplored, is the generation of cointegrates with indigenous plasmids of phytopathogenic *Pseudomonas*. Cointegration, spontaneous or facilitated by prior insertion of Tn7, Tn1, or other transposons in these plasmids, could be directly selected for, in the presence of Km, following conjugation. As in the case of Hfr integration, fusion to an indigenous plasmid would stabilize pAS8Rep-1 in these recipients. Presumptive fusion products have occasionally been detected in stable Km^r transconjugants by gel electrophoresis (this laboratory, unpubl.). These fused derivatives would permit subsequent introduction of indigenous *Pseudomonas* plasmids that, normally, may be unable to transfer or replicate in other hosts, such as *E. coli,* and would facilitate further genetic analysis.

REFERENCES

Bagdasarian, M.; Bagdasarian, M. M.; Coleman, S.; and Timmis, K. N. 1979. New vector plasmids for gene cloning in *Pseudomonas.* In *Plasmids of medical, environmental and commercial importance,* ed. K. N. Timmis and A. Pühler, pp. 411–22. Amsterdam: Elsevier/North-Holland Biomedical Press.

Barth, P. T. 1979a. Plasmid RP4, with *Escherichia coli* DNA inserted *in vitro,* mediates chromosomal transfer. *Plasmid* 2:130–136.

——. 1979b. RP4 and R300B as wide host-range plasmid cloning vehicles. In *Plasmids of medical, environmental and commercial importance,* ed. K. N. Timmis and A. Pühler, pp. 399–410. Amsterdam: Elsevier/North-Holland Biomedical Press.

Barth, P. T.; Datta, N.; Hedges, R. W.; and Grinter, N. J. 1976. Transposition of a deoxyribonucleic acid sequence encoding trimethoprim and streptomycin resistance from R483 to other replicons. *J. Bacteriol.* 125:800–10.

Barth, P. T., and Grinter, N. J. 1974. Comparison of the deoxyribonucleic acid molecular weights and homologies of plasmids conferring linked resistance to streptomycin and sulphonamides. *J. Bacteriol.* 120:618–30.

Beringer, J. E.; Beynon, J. L.; Buchanan-Wollaston, A. V.; and Johnson, A. W. B. 1978. Transfer of the drug resistance transposon Tn5 to *Rhizobium. Nature* (London) 276:633–34.

Boistard, P., and Boucher, C. 1979. P1 group plasmids as tools for genetic study of phytopathogenic bacteria. *Proceedings of the IV International Conference on Plant Pathogenic Bacteria, Angers, France,* 27 August to 2 September 1978, 1: 17–30.

Bolivar, F.; Rodriguez, R. L.; Greene, P. J.; Metlach, M. C.; Heyneker, H. L.; Boyer, H. W.; Crosa, J.; and Falkow, S. 1977. Construction and characterization of new cloning vehicles. II. A multipurpose cloning system. *Gene* 2: 95–113.

Boucher, C.; Barate de Bertalmio, M.; and Dénarié, J. 1977. Introduction of bacteriophage Mu-1 into *Pseudomonas solanacearum* and *Rhizobium meliloti* using R factor RP4. *J. Gen. Microbiol.* 98:253–63.

Boucher, C. A., and Sequeira, L. 1978. Evidence for the cotransfer of genetic markers in *Pseudomonas solanacearum. Can. J. Microbiol.* 24:69–72.

Burkardt, H. -J.; Reiss, G.; and Pühler, A. 1979. Relationship of group R1 plasmids revealed by heteroduplex experiments: RP1, RP4, and R68 and RK2 are identical. *J. Gen. Microbiol.* 114:341–48.

Campbell, A.; Berg, D. E.; Botstein, D.; Lederberg, E. M.; Novick, R. P.; Starlinger, P.; and Szybalski, W. 1979. Nomenclature of transposable elements in procaryotes. *Gene* 6:197–206.

Chakrabarty, A. M.; Myloroie, J. R.; Friello, D. A.; and Vacca, J. G. 1975. Transformation in *Pseudomonas putida* and *Escherichia coli* with plasmid-linked drug-resistance factor DNA. *Proc. Nat. Acad. Sci. (USA)* 72:3647-51.

Chang, A. C. Y., and Cohen, S. N. 1978. Construction and characterization of amplifiable multicopy DNA cloning vehicles derived from the P15A cryptic miniplasmid. *J. Bacteriol.* 134:1141-56.

Comai, L., and Kosuge, T. 1980. Involvement of a plasmid deoxyribonucleic acid in indoleacetic acid production by *Pseudomonas savastanoi*. *J. Bacteriol.* 143: 950-57.

Coplin, D. L.; Sequeira, L.; and Hanson, R. S. 1974. *Pseudomonas solanacearum*: virulence of biochemical mutants. *Can. J. Microbiol.* 20:519-29.

Curiale, M. S., and Mills, D. 1977. Detection and characterization of plasmids in *Pseudomonas glycinea*. *J. Bacteriol.* 131:224-28.

Danilevich, V. N.; Stephanshin, Y. G.; Volozhantsev, N. V.; and Golub, E. I. 1978. Transposon-mediated insertion of R factor into bacterial chromosome. *Molec. Gen. Genet.* 161:337-39.

Day, P. R., and Dodds, A. J. 1979. Viruses of plant pathogenic fungi. In *Viruses and plasmids in fungi*, ed. P. A. Lemke, pp. 201-38. New York: Marcel Dekker.

Dénarié, J.; Rosenberg, C.; Bergeron, B.; Boucher, C.; Michel, M.; and Barate de Bertalmio, M. 1977. Potential of RP4::Mu plasmids for *in vivo* genetic engineering in gram-negative bacteria. In *DNA insertion elements, plasmids and episomes*, ed. A. I. Bukhari, J. A. Shapiro, and S. L. Adhya, pp. 507-20. Cold Spring Harbor, N.Y.: Cold Spring Harbor Laboratory.

Ditta, G.; Stanfield, S.; Corbin, D.; and Helinski, D. R. 1980. Broad host range DNA cloning system for gram-negative bacteria: construction of a gene bank of *Rhizobium meliloti*. *Proc. Nat. Acad. Sci. (USA)* 77:7347-51.

Durbin, R. D., ed. In press. *Toxins in plant disease*. New York: Academic Press.

Fulbright, D. W., and Leary, J. V. 1978. Linkage analysis of *Pseudomonas glycinea*. *J. Bacteriol.* 136:497-500.

Gantotti, B. V.; Patil, S. S.; and Mandel, M. 1979a. Apparent involvement of a plasmid in phaseotoxin production by *Pseudomonas phaseolicola*. *Appl. Environ. Bacteriol.* 37:511-16.

——. 1979b. Genetic transformation of *Pseudomonas phaseolicola* by R-factor DNA. *Molec. Gen. Genet.* 174:101-03.

Gonzalez, C. F., and Vidaver, A. K. 1979a. Analysis of plasmids of syringomycin producing strains of *Pseudomonas syringae*. *Proceedings of the International Conference on Plant Pathogenic Bacteria*, Angers, France, 27 August to 2 September 1978, 1:31-38.

——. 1979b. Bacteriocin, plasmid and pectolytic diversity in *Pseudomonas cepacia* of clinical and plant origin. *J. Gen. Microbiol.* 110:161-70.

——. 1979c. Syringomycin production and holcus spot disease of maize: plasmid associated properties in *Pseudomonas syringae*. *Curr. Microbiol.* 2:75-80.

Grinter, N. J., and Barth, P. T. 1976. Characterization of SmSu plasmids by restriction endonuclease cleavage and compatibility testing. *J. Bacteriol.* 128:394-400.

Gross, D. C., and Vidaver, A. K. 1981. Transformation of *Pseudomonas syringae* by plasmid DNA. *Phytopathology* 1:221. (Abstract)

Guerry, P.; Embden, J. B.; and Falkow, S. 1974. Molecular nature of two noncon-jugative plasmids carrying drug resistance genes. *J. Bacteriol.* 117:619–30.

Hedges, R. W.; Cresswell, J. B.; and Jacob, A. E. 1976. A non-transmissible variant of RP4 suitable as cloning vehicle for genetic engineering. *Fed. Eur. Biol. Soc. Lett.* 61:186–88.

Hedges, R. W., and Jacob, A. E. 1974. Transposition of ampicillin resistance from RP4 to other replicons. *Molec. Gen. Genet.* 132:31–40.

Helinski, D. R. 1977. Plasmid vectors for gene cloning. In *Genetic engineering for nitrogen fixation*, ed. A. Holleander, pp. 19–49. New York: Plenum.

Hershfield, V.; Boyer, H. W.; Yanofsky, C.; Lovett, M. A.; and Helinski, D. R. 1974. Plasmid ColE1 as a molecular vehicle for cloning and amplification of DNA. *Proc. Nat. Acad. Sci. (USA)* 71:3445–50.

Holloway, B.; Krishnapilli, V.; and Morgan, A. F. 1979. Chromosomal genetics of *Pseudomonas*. *Microbiol. Rev.* 43:73–102.

Jacob, A. E.; Cresswell, J. M.; and Hedges, R. W. 1976. Properties of plasmids con-structed by *in vitro* insertion of DNA from *Rhizobium leguminosarum* and *proteus mirabilis* into RP4. *Molec. Gen. Genet.* 147:315–23.

Jacob, A. E., and Grinter, N. J. 1975. Plasmid RP4 as a vector replicon in genetic engineering. *Nature* (London) 255:504–6.

Khan, M.; Kolter, R.; Thomas, C.; Figurski, D.; Meyer, R.; Remaut, E.; and Helin-ski, D. R. 1979. Plasmid cloning vehicles derived from ColE1, F, R6K and RK2. In *Methods in enzymology*, ed. S. P. Collowick and N. O. Kaplan, vol. 68, pp. 268–80. New York: Academic Press.

Kleckner, N.; Roth, J.; and Botstein, D. 1977. Genetic engineering *in vivo* using transposable drug-resistance elements. New methods in bacterial genetics. *J. Mol. Biol.* 116:125–59.

Kosuge, T.; Haeskett, M. G.; and Wilson, E. E. 1966. Microbial synthesis and degra-dation of indole-3-acetic acid. I. The conversion of L-tryptophan to indole-3-acetamide by an enzyme system from *Pseudomonas savastanoi*. *J. Biol. Chem.* 241:3738–44.

Kretschmer, P. J., and Cohen, S. N. 1977. Selected translocation of plasmid genes: frequency and regional specificity of translocation of the Tn3 element. *J. Bacteriol.* 130:888–99.

Lacy, G. H., and Leary, J. V. 1979. Genetic systems in phytopathogenic bacteria. *Ann. Rev. Phytopathol.* 17:181–202.

Lê, T. K. T.; Leccas, D.; and Boucher, C. 1979. Transformation of *Pseudomonas solanacearum* strain K60. *Proceedings of the IV International Conference on Plant Pathogenic Bacteria*, Angers, France, 27 August to 2 September 1978, 2:819–22.

Lindow, S. 1977. Leaf surface bacterial ice nuclei as incitants of frost damage to corn (*Zea mays* L.) and other plants. Ph.D. dissertation, University of Wisconsin-Madison.

Mercer, A. A., and Loutit, J. S. 1979. Transformation and transfection of *Pseudo-monas aeruginosa*. Effects of metal ions. *J. Bacteriol.* 140:37–42.

Meyer, R.; Figurski, D.; and Helsinki, D. R. 1975. Molecular vehicle properties of the broad host range plasmid RK2. *Science* 190:1226–28.

——— . 1977. Properties of the plasmid RK2 as a cloning vehicle. In *DNA insertion elements, plasmids and episomes*, ed. A. I. Bukhari, J. A. Shapiro, and S. L. Adhya, pp. 559–566. Cold Spring Harbor, N.Y.: Cold Spring Harbor Laboratory.

Mindich, L.; Cohen, J.; and Weisburd, M. 1976. Isolation of nonsense suppressor mutants in *Pseudomonas*. *J. Bacteriol.* 126:177–82.

Moore, L.; Warren, G.; and Strobel, G. 1979. Involvement of a plasmid in the hairy root-disease of plants caused by *Agrobacterium rhizogenes*. *Plasmid* 2:617–26.

Nagahari, K., and Sakaguchi, K. 1978. RSF1010 plasmid as a potential cloning vector in *Pseudomonas* species. *J. Bacteriol.* 133:1527–29.

Nagahari, K.; Tanaka, K.; Hisinuma, F.; Kuroda, M.; and Sakaguchi K. 1977. Control of tryptophan synthetase amplified by varying the numbers of composite plasmids in *Escherichia coli* cells. *Gene* 1:141–52.

Panopoulos, N. J. 1979. Plasmid RSF1010, a potential cloning vector in *Pseudomonas phaseolicola*. *Phytopathology* 69:1041. (Abstract)

Panopoulos, N. J.; Guimaraes, W. V.; Hua, S. -S.; Sabersky-Lehman, C.; Resnik S.; Lai, M.; ans Shaffer, S. 1978. Plasmids in phytopathogenic bacteria. In *Microbiology — 1978*, ed. D. Schlessinger, pp. 238–41. Washington, D.C.: American Society for Microbiology.

Panopoulos, N. J.; Sandlin, D. E.; Staskawicz, B. J.; Sato, M.; Peters, S.; Homna, M.; and Orser, C. In press. Tools for *in vivo* and *in vitro* genetic manipulation of phytopathogenic pseudomonads and other non-enteric gram negative bacteria. *Recombinant DNA*. (Abstract)

Panopoulos, N. J., and Schroth, M. N. 1974. Role of flagellar motility in the invasion of bean leaves by *Pseudomonas phaseolicola*. *Phytopathology* 64: 1389–97.

Panopoulos, N. J., and Staskawicz, B. J. Forthcoming. Genetics of production. In *Toxins in plant disease*, ed. R. Durbin. New York: Academic Press.

Panopoulos, N. J.; Staskawicz, B. J.; and Sandlin, D. E. 1979. Search for plasmid-associated properties and for a cloning vector in *Pseudomonas phaseolicola*. In *Plasmids of medical, environmental and commercial importance*, ed. K. N. Timmis and A. Pühler, pp. 365–71. Amsterdam: Elsevier/North-Holland Biomedical Press.

Sakanyan, V. A.; Yakubov, L. Z.; Alikhanian, S. I.; and Stepanov, A. I. 1978. Mapping of RP4 plasmid using deletion mutants of pAS8 hybrid (RP4-ColE1). *Molec. Gen. Genet.* 165:331–41.

Sands, D. C.; Warren, G.; Myers, D. F.; and Scharen, A. L. 1979. Plasmids as virulence markers in *Pseudomonas syringae*. *Proceedings of the IV International Conference on Plant Pathogenic Bacteria*, Angers, France, 27 August to 2 September 1978, 1:39–45.

Sasakawa, C., and Yoshikawa, M. 1980. Transposon (Tn5)-mediated suppressive integration of ColE1 derivatives into the chromosome of *Escherichia coli* K12 (*guaA*). *Biochem. Biophys. Res. Commun.* 96:1364–70.

Sato, M.; Staskawicz, B. J.; and Panopoulos, N. J. Forthcoming. A hybrid plasmid suitable for transposon insertions in phytopathogenic *Pseudomonas*. *Plasmid*.

Staskawicz, B. J.; Sato, M.; and Panopoulos, N. J. 1981. Genetic and molecular

characterizatin of an indigenous conjugative plasmid in the bean wild fire
strain of *Pseudomonas syringae* pv. *tabaci. Phytopathology* 71:257. (Abstract)

Templeton, G. E.; TeBeest, D. O.; and Smith, R. J. 1979. Biological weed control
with mycoherbicides. *Ann. Rev. Phytopathol.* 17:301–10.

Twiddy, W., and Liu, S. C. Y. 1972. Intraspecific transformation of *Pseudomonas
syringae. Phytopathology* 62:794. (Abstract)

Van Vliet, F.; Silva, B.; Van Montague, M.; and Schell, J. 1978. Transfer of RP4::Mu
plasmids to *Agrobacterium tumefaciens. Plasmid* 1:446–55.

Vasile'va, S. V., and D'Yakov, Y. T. 1970. Genetics of phytopathogenic *Pseudomonas* spp. V. Transformation of prototrophicity and virulence in auxotrophic mutants of *P. tabaci* and *P. phaseolicola. Biol. Nauki* 13:103–6. (Summary in *Rev. Plant Pathol.* 50(1971):283)

13

CLONING, TRANSFORMATION, AND ANALYSIS OF YEAST GENES

Jeffrey N. Strathern

INTRODUCTION

In the past decade technical advances have allowed the analysis of the regulation of gene expression at the molecular level. The determination of how a gene is turned on or off and what steps are required between transcription and the final product requires the purification of the gene, its primary and mature transcript, the initial and processed protein (if applicable), and any regulators of the process. This analysis is dependent upon the isolation and amplification of specific deoxyribonucleic acid (DNA) sequences, that is, cloning. In this fashion it has been shown that positive and negative transcriptional controls in prokaryotes can be mediated by site-specific DNA-protein interactions. These same technical advances have provided solutions to several genetic paradoxes by allowing the demonstration of additional mechanisms of regulation. In many cases establishing the biological activity of the cloned fragments requires reintroducing the DNA into the starting organism by transformation or transduction. In this chapter I will review the various technologies that allow cloning and characterization of specific genes as they have been applied to yeast. Further, I will attempt to illustrate how these experiments contributed to understanding the regulation of the yeast genome. It is hoped that such a description will provide, by analogy, a guide to the molecular analysis of regulation in more complex eukaryotes.

There are two separate approaches that can be used to isolate specific genes: screen cloned DNAs for homology to a ribonucleic acid (RNA) or

I would like to acknowledge my colleagues James Hicks, Amar Klar, James Broach, and Kim Nasmyth for their roles in making these molecular approaches to genetics a part of my research. In addition, my thanks to Louisa Dalessandro for preparation of the manuscript and Mike Ockler for Fig. 13.1.

related DNA and screen or select for biological function. Each approach has its strengths and weaknesses.

ISOLATION OF CLONED GENES BY SEQUENCE HOMOLOGY

Radioactively labeled RNA or DNA can be used to screen clones for those that contain homologous sequences. For DNAs cloned into bacterial plasmids, bacterial colonies are replicated onto nitrocellulose and then lysed. The DNA from the lysed colonies binds to the nitrocellulose and can be screened for homology to the probe as described by Grunstein and Hogness (1975). Colony screening is applicable to unique genes from organisms with genomes about the size of yeast (14,000 kilobases) or smaller. For DNA cloned into lambdoid phages the DNA in the plaques is transferred onto nitrocellulose and screened as described by Benton and Davis (1977). Because thousands of plaques can be screened on each petri plate, plaque screening has been successfully applied to several higher eukaryotes.

Ribosomal RNAs

Ribosomal RNAs (rRNAs) can be fairly readily purified and radioactively labeled for use as probes. Further, in most organisms the rDNA genes are reiterated. Hence, they are among the easiest genes to clone and have been isolated from many organisms. In yeast there are about 120 tandemly repeated copies of an 8.4 kb sequence that contains the 25s, 18s, 5.8s, and 5s rRNA genes (Cramer, Farrally, and Rownd 1976). A plasmid containing this unit ws isolated by preselecting plasmids that had the correct size insert and then screening for homology to [^{125}I]-labeled rRNA (Bell et al. 1977). Using this or similar clones the pattern of transcription of this repeat unit has been established (Bell et al. 1977; Kramer, Philippsen, and Davis 1978), and the pathway of RNA maturation has been determined (reviewed by Warner [1981]).

The genetic map position of the rRNA genes was the subject of some controversy, which was finally resolved when the copy number was shown to be proportional to the number of copies of chromosome XII (Petes 1979). The interesting question of how tandem repeats are kept uniform was approached by reintroducing a modified rRNA gene by transformation (see below). Unequal crossing-over between sister chromatids during mitosis occurs at a rate sufficient to maintain uniformity (Petes 1979; Szostak and Wu 1980). Thus the cloning of rRNA genes in yeast contributed to understanding how ribosomes are constructed, how reiterated genes are coordinately regulated, and how tandem arrays are generated and stabilized.

Transfer RNA

Nonspecific transfer RNA (tRNA) probes can be made by growing yeast on [^{32}P]-orthophosphate and then isolating the 4s portion of a sucrose gradient. Using such a probe to screen a collection of 4,000 colonies containing yeast *Hind*III restriction fragments cloned onto a plasmid, Abelson and his colleagues identified 175 clones with tRNA genes (Beckman, Johnson, and Abelson 1977). Their analysis indicated that the tRNA genes are not clustered. Specific tRNA genes were identified among those clones by hybridization to radioactively labeled purified tRNA (Beckman, Johnson, and Abelson 1977).

Further, specific tRNA genes have been isolated directly by screening of yeast-λ hybrid plaques using purified tRNA as the probe (Olson, Loughney, and Hall 1979). In several cases these cloned tRNA genes could be correlated with tRNA genes identified by nonsense suppressor mutations. Analysis of these cloned genes led to the demonstration that some tRNA genes have an intervening sequence near the anticodon (Goodman, Olson, and Hall 1977; Valenzuela et al. 1978) and defined a new step in the maturation of tRNAs in which those extra 14 to 60 bases are removed (Knapp et al. 1979).

Synthetic Probes

The sequence of a protein allows a prediction of the DNA sequence that codes it. Because of the degeneracy of the genetic code, such predictions are generally not particularly useful unless combined with the information obtained by sequencing the corresponding proteins made by frameshift mutations. From this type of analysis Stewart and Sherman (1973) predicted the first 44 nucleotides of the gene for iso-1-cytochrome c. By using 13-base-pair (Montgomery et al. 1978) and 15-base-pair (Szostak et al. 1979) synthetic probes the genomic sequence coding for iso-1-cytochrome c has been identified in restriction digests of total yeast DNA and cloned from λ-yeast hybrid banks. The isolation of the gene for iso-1-cytochrome c(CYC1) – and by homology, iso-2-cytochrome c(CYC7) – was important because it allowed the physical analysis of a large collection of interesting mutations including mutations that effect the regulation of these genes. However, the synthetic-probe approach has obvious shortcomings as a general approach.

Partially Purified Messenger RNA

In some cases it is possible to physically enrich for a particular messenger RNA (mRNA). For example, Holland and Holland (1978) size-fractionated yeast mRNA by formamide-polyacrylamide gel electrophoresis. The different size fractions were electroeluted and translated in a wheat germ cell-free extract. This process allowed the identification of the various size fractions containing the glycolytic enzymes enolase, glyceraldehyde-3-phosphate dehydrogenase and phosphoglycerate kinase. These partially purified fractions were used to make complementary DNA (cDNA) probes, which were in turn used to screen plasmid clones of yeast DNA (Holland, Holland, and Jackson 1979). They were able to use the cDNA from the partially purified mRNA as a probe to restriction digests of whole genomic DNA using the technique of Southern (1975).

Having identified the sizes of the interesting restriction fragments, Holland, Holland, and Jackson (1979) were able to facilitate their screen by restricting the plasmid bank to hybrid molecules of the correct size. By using the partially purified probe to screen the correct size-fractionated plasmid bank, they were able to identify a plasmid clone containing the glyceraldehyde-3-phosphate dehydrogenase gene, as well as separate clones for the enolase and phosphoglycerate kinase genes.

Differentially Induced mRNAs

The spectrum of mRNAs present in cells changes as cells respond to their environment or undergo differentiation. Thus, when mRNAs from cells grown under different conditions are compared as probes with clones, it is possible to identify clones carrying differentially expressed genes. For example, Hopper and colleagues (Hopper, Broach, and Rowe 1978; Hopper and Rowe 1978) demonstrated that translatable messages corresponding to the genes for galactose utilization in yeast GAL1 and GAL7 were present in cells induced for growth on galactose but reduced or absent in cells grown on glucose or acetate.

St. John and Davis (1979) modified the basic plaque screening technique of Benton and Davis (1977) so that they had duplicates of each filter. One filter was hybridized to a probe made from mRNA of glucose-grown cells, while the other filter was hybridized to a probe derived from galactose-grown cells. In addition to clones that hybridized to both probes or to neither probe, these filters identified clones that hybridized to one probe but not to the other. About 1 percent of the clones showed specific hybridization to the probe derived from galactose-induced cells. This analysis allowed the isolation and identification of several of the genes of the galactose utilization pathway. These clones then opened the doorway to understanding the molecular basis of the regulation of these galactose genes in concert with the considerable genetics of galactose regulation (Douglas and Hawthorne 1966; Matsumoto, Toh-e, and Oshima 1978).

St. John and Davis (1979) have an excellent discussion of the technical limitations of this differential screening technique. It is obviously very sensitive to the complexity of the probe and the fraction of the probe corresponding to differentially expressed mRNA. For relatively abundant messages that are specific to response to an environment or state of differentiation, this technique can be applied to higher eukaryotes (S. Hughes 1980: personal communication). For messages that are not abundant, the high complexity of the probe results in a high ratio of plaques that give positive signals under both conditions being tested and necessitates reducing the number of plaques per filter being screened to the point where effectively screening a higher eukaryote becomes quite tedious. For many cell-type-specific or inducible genes this difficulty can be overcome by combining size fractionation of the mRNA (to reduce the complexity) with the differential hybridization screen.

ISOLATION OF CLONED GENES BY BIOLOGICAL FUNCTION

Screening or selecting for function requires a means of introducing the cloned DNA into an organism. In prokaryotes transducing phage and transformation allow the fairly efficient shuffling of DNAs. No virus capable of infecting yeast — and hence of potential use as a vector — has been isolated. Further, intact yeast are not readily transformed by DNA. It is instructive to trace the origins of efficient transformation in yeast as an analogy for those higher eukaryotes that have no natural vector and/or as yet no good selectable cloned marker to identify transformants.

Attempts to transform yeast with total yeast DNA selecting for complementation of a nutritional marker showed some minimal success (Oppenoorth 1960) but were trying to solve three difficult problems at once. These pioneering experiments had the selectable sequence present in low concentration because the complexity of the DNA is large. Further, the conditions under which yeast would take up DNA and incorporate it into its genome could only be guessed by analogy with transformation in prokaryotes and mammalian tissue culture cells. Finally, the transformants were very difficult to distinguish from revertants.

Selection of Yeast Genes in Prokaryotes

The important breakthrough involved the selection of yeast genes to function in prokaryotes. Struhl, Cameron, and Davis (1976) constructed a pool of λ-yeast hybrid phage. These phage were infected into a *hisB* (imidazole glycerol phosphate (IGP) dehydratase) *Escherichia coli* strain under conditions that allowed the stable integration of the phage into the bacterial genome. These lysogens were then selected for the His⁺ phenotype. Survivors were found at a frequency of 10^{-8}, although the *hisB* mutation itself is nonreverting. These survivors contained λ-yeast hybrids containing a unique yeast DNA fragment that was subsequently shown to contain the yeast *HIS3* gene (IGP dehydratase) (Scherer and Davis 1979). Ratzkin and Carbon (1977) constructed a bank of ColE1-yeast hybrid plasmids that could be introduced into *E. coli* by transformation and selected by resistance to colicin E1. This pool was screened for the ability to complement *leuB* (β-isopropylmatate dehydrogenase) and *hisB* mutations in *E. coli* and yielded the plasmid clones pYeleu10 and pYehis, which were subsequently shown to contain the yeast *LEU2* and *HIS3* genes, respectively (Hicks and Fink 1977). Thus, by selection for function in a heterologous system (yeast genes in *E. coli*) Struhl, Cameron, and Davis (1976) and Ratzkin and Carbon (1977) obtained purified and amplified yeast genes with which to work out the mechanics of efficient yeast transformation. Several other yeast genes have been isolated by selecting for complementation of corresponding bacterial mutation with a success ratio of about 25 percent (Clarke and Carbon 1978).

Immunological Screening

Clarke, Hitzeman, and Carbon (1979) have described a useful variation on selecting for function of yeast genes in *E. coli* based on immunological detection. This procedure has lower requirements for the proper transcription processing and translation of eukaryotic genes in *E. coli*. Further, it can allow the detection of clones carrying genes that are not selectable in *E. coli* or that cannot function in *E. coli* such as the gene for one subunit of complex enzyme. Each clone is screened for the desired antigen by exposing it first to immobilized antibody (antibody bound to plastic, or CNBr-activated paper); then to soluble radioactive antibody. This sandwich-labeling technique can be applied to pools of clones using plastic wells or individual lysed colonies transferred to the antibody-CNBr paper. Using these techniques the gene for 3-phosphoglycerate kinase *(PKG)* has been isolated from yeast (Hitzeman, Clarke, and Carbon 1980). This gene is of particular interest both because it maps near a centromere and because it is very efficiently transcribed and translated, yielding several percent of the total cell protein.

The detection of eukaryotic genes in *E. coli* by antibody screening has the advantages described above but still is plagued by the difficulties inherent in screening large genomes and, of course, requires the purification of the desired protein and an antibody directed toward it. Nonetheless, it should prove quite useful for the cloning of genes for which there are no mutations and for which the mRNA is difficult to purify but for which the protein can be isolated.

Selection of Yeast Genes in Yeast

Hinnen, Hicks, and Fink (1978) exposed yeast that had been enzymatically stripped of their cell walls to the pYeleu10 clone in the presence of polyethylene glycol calcium and an osmotic stabilizer. Under these conditions yeast cells are known to fuse (van Solingen and van der Plaat 1977). About 1 per 10^7 of the regenerated spheroplasts had the Leu$^+$ phenotype. These Leu$^+$ yeast had the ColE1 plasmid integrated into the yeast DNA in a single copy where it behaved as a normal Mendelian gene. Demonstration of the integrated ColE1 plasmid in the transformants provided a single molecular proof of transformation.

The majority of the Leu$^+$ yeast transformants are duplications of the *LEU2* gene on chromosome III caused by integration of the hybrid plasmid by homologous recombination. In fact, in general most or all of the transformants of yeast using DNAs that are not capable of replicating as plasmids (see below) apparently involve homologous recombination (Hicks, Hinnen, and Fink 1979). This feature is in contrast to DNA transformation in mammalian tissue culture lines and should be appreciated as a powerful genetic tool, as discussed below.

The next step in the evolution of transformation in yeast involved the observation that some hybrid bacterial plasmid-yeast DNA clones transformed at significantly higher frequencies and that the transformed DNAs were maintained as plasmids in the yeast cells (Fig. 13.1). Such plasmids obviously had a sequence that could function as an origin of DNA replication. It became clear that with the inclusion of a yeast replication origin and a selectable yeast gene useful yeast cloning vectors could be developed (Beggs 1978; Broach, Strathern, and Hicks 1979; Struhl et al. 1979). These studies resulted in vectors that were selectable and that could replicate in bacteria (so that hybrid banks could be made and individual clones amplified) and that were selectable and could replicate in yeast (so that transformation was efficient and biological function in yeast could be tested). Using such vectors the transformation rate could be improved to about 1 per 10^4 regenerated spheroplasts. Under these conditions it is possible to make a bank of hybrid yeast molecules and screen yeast transformants for function of genes one would not expect to function in yeast.

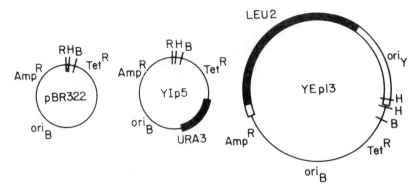

FIG. 13.1. Three type of plasmids for cloning yeast genes. pBR322 (Bolivar et al. 1977) replicates in *E. coli* and has markers selectable in *E. coli* and resistance to ampicillin (Amp[R]) and tetracycline (TET[R]). There are several unique endonuclease cleavage sites in pBR322 that can be used as cloning sites including *Hind*III (**H**), *Eco*RI (**R**), and *Bam*HI (**B**). pBR322 is useful for generation of hybrid DNA and banks for colony hybridization screening with a homologous probe, for immunological screening (see text), or for selection for complementation of *E. coli* mutations. YIp5 (Struhl et al. 1979) is a 1.1 kilobase insert containing the *URA3* gene from yeast inserted into pBR322. YIp5 replicates in bacteria but not in yeast. It is particularly useful for the identification of yeast autonomously replicating sequences, for the identification of cloned sequences by the analysis of integrative transformants (see text), and for gene replacement or marker rescue experiments. YEp13 (Broach, Strathern, and Hicks 1979) replicates in yeast and in *E. coli* and is selectable in either organism. The yeast origin of replication is derived from a natural plasmid found in yeast, or the 2 micron circle. Transformation with YEp13 is much more efficient than with YIp5. YEp13 can be used as a *Bam*HI insertion cloning vector or a *Hind*III replacement vector. YEp13 is found in about 20 copies per yeast cell but segregates cells that do not contain the plasmid. YEp13 is useful for the screening or selecting of cloned genes that function in yeast.

GENETIC IDENTIFICATION OF CLONED DNAs

The observation that a cloned sequence complements a genetic defect in yeast or an analogous defect in *E. coli* is not sufficient to identify the clone as containing the desired gene. For example, Ratzkin and Carbon (1977) isolated several yeast sequences that were able to suppress the *leuB* mutation in *E. coli*. Although one turned out to be the analogous *LEU2* gene from yeast, the basis of the suppression of *leuB* by the other yeast genes is not understood. Several techniques have been developed to determine whether a cloned sequence corresponds to the desired gene. Hicks and Fink (1977) compared the amount of hybridization of the clone pYeleu10 with DNA from several yeast strains with chromosome III in different ratios to the other chromosomes. They found that the amount of hybridization was proportional to the number of chromosome III copies as expected if pYeleu10 carried the yeast *LEU2* gene from chromosome III.

The fact that unrelated yeast strains frequently have different restriction endonuclease cleavage maps surrounding any given cloned sequence provides a second means of identifying the genetic location of a cloned DNA. For example, eight locations in the yeast genome that code for tyrosine tRNA had been identified by their ability to mutate to ochre suppressors (Hawthorne and Leupold 1974). These tyrosine tRNA hybridize to eight places in the genome — that is, to eight bands in a restriction digest of yeast DNA. By screening a variety of yeast strains, restriction endonuclease cleavage fragment polymorphisms were identified for these bands. By correlating the segregation pattern of the suppressor genes with the segregation patterns of the band polymorphisms, it is possible to identify the genetic location of the bands (Olson, Loughney, and Hall 1979). Fig. 13.2 demonstrates this kind of analysis for a clone of the *CAN1* gene from yeast (described below).

The fact that transformation of yeast by nonreplicating plasmids occurs by homologous recombination provides a third way of confirming the identity of cloned genes. Hinnen, Hicks, and Fink (1978) determined that Leu$^+$ transformants using the clone pYeleu10 resulted in the integration of the bacterial plasmid into the yeast genome. The bacterial plasmid segregated as a chromosomal marker and could be scored by hybridization to yeast colonies lysed onto nitrocellulose. The observation that the segregation showed tight linkage to the *LEU2* gene again confirmed pYeleu10 as a clone of that yeast gene.

Transformants formed by homologous recombination of the clone into the chromosome have a duplication of cloned gene flanking the vector. This arrangement allows either the isolation of the chromosomal allele onto the plasmid or the replacement of the chromosomal allele by the DNA from the clone. To isolate the genomic allele, DNA from the transformed strain is cut with a restriction endonuclease that does not cut into the clone and then cyclized and transformed into bacteria selecting for the plasmid. The resultant clone contains both the original cloned DNA and the genomic sequence. More elegant variations on this theme are possible, depending on the nature of the original clone. For example, an *EcoRI-HindIII* fragment in YIp5 will integrate into the genome so that an *EcoRI* digest and cyclization will yield the DNA on one side of the plasmid, while a *HindIII* digest and cyclization yields the other side. One of these two should have the genomic allele. J. Broach (1980: personal communication) has used this approach to take a clone for a wild type tRNA gene and isolate the suppressor allele.

The ability to specifically replace genomic alleles with cloned genes opens the way to the functional analysis of in vitro made deletion, inser-

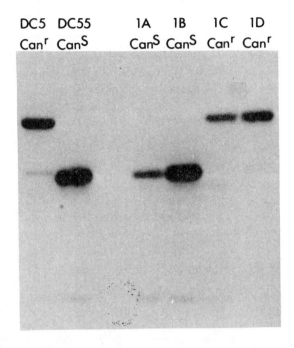

FIG. 13.2. Linkage of the fragment cloned in TLC-1 to the *CAN1* gene. DNA was isolated from strain DC5 (lane 1) and strain DC55 (lane 2), and the four spore clones (lanes 3 through 6) were derived from dissection of one ascus obtained after mating strains DC5 and DC55 and sporulation of the resultant diploid. 5 μg of each DNA were digested with *Eco*RI, fractionated on a 1 percent agarose gel, transferred to nitrocellulose, and hybridized with labeled TLC-1-RI DNA. The autoradiogram in the figure shows the size of the restriction fragment in the genome of each strain that is homologous to TLC-1-RI DNA. Strains DC5, 1C, and 1D are *can1⁻* (Can^r), and strains DC55, 1A, and 1B are *CAN1⁺* (Can^s). Reprinted from *Gene* 8 [1979]: 121–33, courtesy of Elsevier/North-Holland Biomedical Press, Amsterdam.

tion, or substitution variants. The requirement that this analysis take place in the absence of the native allele is difficult to meet in those systems where transformation does not involve homologous recombination. However, in yeast the tandem duplication that results from the integration of a nonreplicative vector is not very stable. The vector-plus-insert tends to recombine out. In some of these cells the recombination results in the net replacement of the native allele with the variant sequence. Scherer and Davis (1979) used this technique to replace the wild type *HIS3* gene in yeast with a *his3⁻* deletion that had been made in vitro. The functional analysis of in vitro made mutations will be particularly useful in defining sequences involved in gene regulation.

EXAMPLES

Three examples are described below to illustrate how these various techniques have been combined to allow a molecular analysis of the genetics and biology of *Saccharomyces*.

Arginine Permease

Many proteins are part of complexes or undergo processing that can only occur in the native organism. The arginine permease of yeast is a good candidate for such a protein. In addition to its intrinsically interesting features as a eukaryotic transport system, the gene for the arginine permease *(CAN1)* has been the subject of considerable genetic investigation (Whelan, Gocke, and Manney 1979). Mutants in *CAN1* can be readily selected because they make cells unable to transport the toxic amino acid analogue canavanine.

The *CAN1* locus was isolated by screening a bank of total yeast DNA cloned into YEp13 for the ability to transform a *leu2⁻ arg4⁻ can1⁻* yeast strain to growth on minimal medium plus arginine (Broach, Strathern, and Hicks 1979). Such yeast strains were able to supply their arginine requirement from the medium because of the *CAN1* gene on the plasmid. The YEp13:*CAN1* (TLC-1) plasmid was isolated into bacteria by using DNA from the transformed yeast to transform *E. coli* to ampicillin resistance. This cloned yeast DNA was positively identified as the *CAN1* gene by showing that the meiotic segregation of a restriction endonuclease cleavage site polymorphism near the cloned DNA is genetically linked to the *CAN1* locus (Fig. 13.2). Using this clone to isolate the messenger RNA from *CAN1* and translate it, it should be possible to identify the arginine permease and determine what steps are involved in its maturation and localization into the plasma membrane.

Centromeres in Bacterial Systems

A centromere is an excellent example of a eukaryotic "gene" that one would not expect to function in bacteria. Clarke and Carbon (1980) identified a short yeast DNA that causes plasmids in yeast to segregate as chromosomes. By using a *LEU2*-containing plasmid as a probe, a series of overlapping plasmids were obtained that represented 25 kilobases near *LEU2* and the centromere of chromosome III in yeast. These plasmids included the *CDC10* gene, a locus tightly linked to the centromere. Clarke and Carbon observed that plasmids containing this sequence were relatively much more stable mitotically and meiotically than related yeast plasmids. Most strikingly, the plasmid segregates at the first meiotic division just as a centromere-linked trait. The sequence necessary for the property is less than 500 bases long. No evidence for homology to other centromeres was obtained. Alterations made in this sequence can be reintroduced into yeast and assayed for biological function in an attempt to understand the role of the DNA at the centromere in the mechanics of inheritance. Further, this sequence should allow the construction of a yeast cloning vector that has a low copy number but is stable.

MAT Locus

The three cell types in *Saccharomyces* yeast (**a**, α, and **a**/α) are controlled by the *MAT* locus. The codominant *MAT*a and *MAT*α alleles regulate cell type by controlling the expression of cell-type-specific genes throughout the genome. Such a regulatory gene could not be expected to be isolated by function in a prokaryote, and no specific probe for the gene or gene products existed. The *MAT*α gene was isolated by screening a yeast DNA pool in YEp13 for the ability to suppress *mat$^-$* mutations (Hicks, Strathern, and Klar 1979; Nasmyth and Tatchell 1980). When used as probes to restriction digests of total yeast DNA, these clones demonstrated that *MAT*a and *MAT*α differed by a substitution of several hundred base pairs and that cryptic *MAT*-like sequences existed at two other places in the genome.

*MAT*a and the cryptic loci (*HML* and *HMR*) were subsequently isolated by screening plasmid or phage pools of yeast DNA clones for homology. The physical investigation of the *MAT* locus was of keen interest because some strains of yeast (homothallic) are able to switch from *MAT*a to *MAT*α (or from *MAT*α to *MAT*a), and hence to change cell type, very efficiently. Such homothallic strains were able to make these sequence changes at the mating-type locus by making replicas of the cryptic copies and substituting them into *MAT* (Hicks, Strathern, and Herskowicz 1977; Nasmyth and Tatchell 1980; Strathern et al. 1980).

The transposition-mediated activation of cell type regulatory genes is a novel mechanism for cell type differentiation, which, because it involves reversible alterations of the genome, is applicable to higher eukaryotes. Further, the positional effect that allows genes to be expressed at one location in the genome while unexpressed in another location has important analogies in higher eukaryotes. The investigation of this phenomenon in yeast is further facilitated because several recessive mutations that lead to the expression of the cryptic genes have been isolated, suggesting that the silent genes are actively turned off (Haber and George 1979; Klar, Fogel, and MacLeod 1979; Rine et al. 1979). Thus the cloning of the *MAT*, *HML*, and *HMR* loci and the ability to reintroduce these DNAs into yeast opened the door to the physical characterization of a regulatory gene, the physical analysis a specific rearrangement of the genome involved in differentiation, and the investigation of a genetic position effect.

PERSPECTIVES

In many ways the molecular analysis of gene regulation in yeast is still in its infancy. The isolation of a gene is only the beginning of the analysis of its control. Although the initial successes in cloning involved genes for which there were easily isolated probes or which could be isolated by function, the attention quickly shifted to those loci that previous genetic analysis suggested were regulated by cell type, position in the cell cycle, or response to the environment. While several regulated loci have been cloned, their transcripts characterized and even variants isolated with defects in response to that regulation, no trans-acting regulatory molecule has been identified. We can look forward to this level of analysis in the near future.

As the number of yeast origins of replication that have been cloned and analyzed increases, the features defining such sites should become apparent. This knowledge together with the development of yeast-based in vitro systems capable of synthesizing DNA will lead to molecular description of how the genome is replicated.

We can expect the physical analysis of the mechanics of mitosis and meiosis to proceed as additional centromeres are isolated and compared and as cellular components that specifically recognize these centromere sequences are identified. At the same time we can expect that the ends or telomeres of yeast chromosomes will be isolated so that the suggestion that they are cross-linked (Forte and Fangman 1979) can be tested and their unique problems of replication investigated.

Finally, the efficient transformation of yeast together with its excellence as an organism for genetic research suggests that yeast may be of significant use in the isolation and investigation of genes from other eukaryotic organisms.

I made no attempt to describe the exact protocols used in the cloning and characterization of yeast genes because these details may be restricted to yeast and/or are likely to change rapidly. However, an appreciation for the power of this approach should be long-lived and transferable to other organisms.

REFERENCES

Beckman, J.; Johnson, P. F.; and Abelson, J. 1977. Cloning of yeast transfer RNA genes in *Escherichia coli. Science* 196:205–10.

Beggs, J. D. 1978. Transformation of yeast by a replicating hybrid plasmid. *Nature* 275:104–9.

Bell, G. I.; DeGoungro, L. J.; Gelfand, D. H.; Bishop, R. J.; Valenzuela, P.; and Bulter, W. J. 1977. Ribosomal RNA genes of *Saccharomyces cerevisiae*. *J. Biol. Chem.* 252:8118–25.

Benton, W., and Davis, R. 1977. Screening of λgt recombinant clones by hybridization to single plaques in situ. *Science* 196:180–82.

Bolivar, R.; Rodriguez, R. L.; Greene, R. J.; Betlach, M. C.; Heyneker, H. L.; Boyer, H. W.; Crosa, J.; and Falkow, S. 1977. Construction and characterization of new cloning vehicles: II. A multipurpose cloning system. *Gene* 2:95–113.

Broach, J. R.; Strathern, J. N., and Hicks, J. B. 1979. Transformation in yeast: development of a hybrid cloning vector and isolation of the *CAN1* gene. *Gene* 8:121–33.

Clarke, L., and Carbon, J. 1978. Functional expression of cloned yeast DNA in *Escherichia coli*: specific complementation of argininosuccinate lygase (*argH*) mutations. *J. Mol. Biol.* 120:517–32.

——— . 1980. Isolation of a yeast centromere and construction of functional small circular chromosomes. *Nature* (London) 287:504.

Clarke, L.; Hitzeman, R., and Carbon, J. 1979. Selection of specific clones from colony banks by screening with radioactive antibodies. In *Methods in enzymology*, ed. S. P. Collowick and N. O. Kaplan, vol. 68, pp. 436–42. New York: Academic Press. (Edited by R. Wu)

Cramer, J. H.; Farrally, F. W.; and Rownd, R. H. 1976. Restriction endonuclease analysis of ribosomal DNA from *Saccharomyces cerevisiae*. *Molec. Gen. Genet.* 148:233–41.

Douglas, H. C., and Hawthorne, D. C. 1966. Regulation of genes controlling synthesis of the galactose pathway enzymes in yeast. *Genetics* 54:911–16.

Forte, M. A., and Fangman, W. L. 1979. Yeast chromosomal DNA molecules have strands which are crosslinked at their termini. *Chromosoma* 72:131–39.

Goodman, H. M.; Olson, M. R.; and Hall, B. D. 1977. Nucleotide sequence of a mutant eukaryotic gene: the yeast tyrosine-inserting ochre suppressor *SUP4-0*. *Proc. Nat. Acad. Sci. (USA)* 74:5453–57.

Grunstein, M., and Hogness, D. S. 1975. Colony hybridization: a method for the isolation of cloned DNAs that contain a specific gene. *Proc. Nat. Acad. Sci. (USA)* 72:3961–65.

Haber, J. E., and George, J. P. 1979. A mutation that permits the expression of normally silent copies of mating-type information in *Saccharomyces cerevisiae*. *Genetics* 93:13–32.

Hawthorne, D. C., and Leupold, U. 1974. Suppressors in yeast. *Curr. Top. in Microbiol. Immunology* 64:1–47.

Hicks, J. B., and Fink, G. R. 1977. Identification of chromosomal location of yeast DNA from hybrid plasmid pYeleu10. *Nature* 269:265–67.

Hicks, J. B.; Hinnen, A.; and Fink, G. R. 1979. Properties of yeast transformation. *Cold Spring Harbor Symp. Quant. Biol.* 43:1306–13.

Hicks, J. B.; Strathern, J. N.; and Herskowicz. I. 1977. The cassette model of mating-type interconversion. In *DNA insertion elements, plasmids and episomes,* ed. A. I. Bukhari, J. A. Shapiro, and S. L. Adhya, pp. 457–62. Cold Spring Harbor, N.Y.: Cold Spring Harbor Laboratory.

Hicks, J. B.; Strathern, J. N.; and Klar, A. J. S. 1979. Transposable mating type genes in *Saccharomyces cerevisiae*. *Nature* 282:478–82.

Hinnen, A.; Hicks, J. B.; and Fink, G. R. 1978. Transformation in yeast. *Proc. Nat. Acad. Sci. (USA)* 75:1929–33.

Hitzeman, R. A.; Clarke, L.; and Carbon, J. 1980. Isolation and characterization of the yeast 3-phosphoglycerokinase gene *(PGK)* by an immunological screening technique. *J. Biol. Chem.* 255:12073–80.

Holland, M. J., and Holland, J. P. 1978. Isolation and identification of yeast messenger ribonucleic acids coding for enolase, glyceraldehyde-3-phosphage dehydrogenase, and phosphoglycerate kinase. *Biochemistry A.C.S.* 17:4900–7.

Holland, M. J.; Holland, J. P.; and Jackson, K. A. 1979. Cloning of yeast genes coding for glycolytic enzymes. In *Methods in enzymology*, ed. S. P. Collowick and N. O. Kaplan, vol. 68, pp. 408–19. New York: Academic Press. (Edited by R. Wu).

Hopper, J. E.; Broach, J. R.; and Rowe, L. B. 1978. Regulation of the galactose pathway in *Saccharomyces cerevisiae*: induction of uridyl transferase mRNA and dependency on *GAL4* gene function. *Proc. Nat. Acad. Sci. (USA)* 75:2878–82.

Hopper, J. E., and Rowe, L. B. 1978. Molecular expression and regulation of the galactose pathway genes in *Saccharomyces cerevisiae*: distinct messenger RNAs specified by *GAL1* and *GAL7* genes in the *GAL7-GAL10-GAL1* cluster. *J. Biol. Chem.* 253:7566–69.

Klar, A. J. S.; Fogel, S.; and MacLeod, K. 1979. *MAR1* a regulator of *HM*a and *HM*α loci in *Saccharomyces cerevisiae*. *Genetics* 93:37–50.

Knapp, G.; Ogden, R. C.; Peebles, C. L.; and Abelson, J. 1979. Splicing of yeast tRNA precursors: structure of the reaction intermediate. *Cell* 18:37–45.

Kramer, R. A.; Philippsen, P.; and Davis, R. W. 1978. Divergent transcription in the yeast ribosomal RNA coding region as shown by hybridization to separated strands and sequence analysis of cloned DNA. *J. Mol. Biol.* 123:405–16.

Matsumoto, J.; Toh-e, A.; and Oshima, Y. 1978. Genetic control of galactokinase synthesis in *Saccharomyces cerevisiae*: evidence for constitutive expression of the positive regulatory gene *GAL4*. *J. Bacteriol.* 134:446–57.

Montgomery, D. L.; Hall, B. D.; Gillian, S.; and Smith, M. 1978. Identification and isolation of the yeast cytochrome *c* gene. *Cell* 14:673–80.

Nasmyth, K., and Tatchell, K. 1980. The structure of transposable yeast mating type loci. *Cell* 19:753–64.

Olson, M. V.; Loughney, K.; and Hall, B. D. 1979. Identification of the yeast DNA sequences that correspond to specific tyrosine-inserting nonsense suppressor loci. *J. Mol. Biol.* 132:387–410.

Oppenoorth, W. F. F. 1960. Modification of the hereditary character of yeast by ingestion of cell free extracts. Eur. Brewery Convention, Elsevier, Amsterdam, p. 180. (Abstract)

Petes, T. D. 1979. Yeast ribosomal DNA genes are located on chromosome XII. *Proc. Nat. Acad. Sci. (USA)* 76:410–14.

Ratzkin, B., and Carbon, J. 1977. Functional expression of cloned yeast DNA in *Escherichia coli*. *Proc. Nat. Acad. Sci. (USA)* 74:487–91.

Rine, J.; Strathern, J. N.; Hicks, J. B.; and Herskowitz, I. 1979. A suppressor of mating type locus mutations in *Saccharomyces cerevisiae*: evidence for and identification of cryptic mating type loci. *Genetics* 93:877–901.

St. John, T. P., and Davis, R. W. 1979. Isolation of galactose-inducible DNA sequences from *Saccharomyces cerevisiae* by differential plaque filter hybridization. *Cell* 16:443–52.

Scherer, S., and Davis, R. W. 1979. Replacement of chromosome segments with altered DNA sequences constructed *in vitro*. *Proc. Nat. Acad. Sci. (USA)* 76:4951–55.

Southern, E. 1975. Detection of specific DNA sequences among DNA fragments. *J. Mol. Biol.* 98:503–17.

Stewart, J. W., and Sherman, F. 1973. Mutations at the end of the iso-1-*cytochrome* gene of yeast. In *The biochemistry of gene expression in higher organisms*, ed. J. K. Pollack and J. W. Lee, p. 56. Sydney, Australia: Australian and New Zealand Book Co.

Strathern, J. W.; Spatola, E.; McGill, C.; and Hicks, J. B. 1980. Structure and organization of transposable mating type cassettes in *Saccharomyces* yeasts. *Proc. Nat. Acad. Sci. (USA)* 77:2839–43.

Struhl, K.; Cameron, J. R.; and Davis, R. W. 1976. Functional genetic expression of eukaryotic DNA in *Escherichia coli*. *Proc. Nat. Acad. Sci. (USA)* 73:1471–75.

Struhl, K.; Stinchcomb, D. T.; Scherer, S.; and Davis, R. W. 1979. High-frequency transformation of yeast: autonomous replication of hybrid DNA molecules. *Proc. Nat. Acad. Sci. (USA)* 76:1035–39.

Szostak, J. W.; Stiles, J. I.; Tye, B. -K.; Chiu, P.; Sherman, F.; and Wu, R. 1979. Hybridization with synthetic oligonucleotides. In *Methods in enzymology*, ed. S. P. Collowick and N. O. Kaplan, vol. 68, pp. 419–28. New York: Academic Press. (Edited by R. Wu)

Szostak, J. W., and Wu, R. 1980. Unequal crossing over in the ribosomal DNA of *Saccharomyces cerevisiae*. *Nature* (London) 284:426–30.

Valenzuela, P.; Venagas, A.; Weinberg, F.; Bishop, R.; and Rutter, W. J. 1978. Structure of phenylanine-tRNA genes: intervening DNA segment within region coding for tRNA. *Proc. Nat. Acad. Sci. (USA)* 75:190–94.

van Solingen, P., and van der Plaat, J. B. 1977. Fusion of yeast spheroplasts. *J. Bacteriol.* 130:946–47.

Warner, J. F. 1981. The yeast ribosome: structure, function and synthesis. In *Molecular biology of the yeast Saccharomyces*, ed. J. N. Strathern, E. W. Jones, and J. Broach, vol. 1. Cold Spring Harbor, N.Y.: Cold Spring Harbor Press.

Whelan, W. L.; Gocke, E.; and Manney, T. R. 1979. The *CAN1* locus of *Saccharomyces cerevisiae*: fine-structure analysis and forward mutation rates. *Genetics* 91:35–51.

14

A VIRUS VECTOR FOR GENETIC ENGINEERING IN INVERTEBRATES

Lois K. Miller

INTRODUCTION

Excellent prokaryotic host-vector systems are currently available for cloning and propagating portions of eukaryotic deoxyribonucleic acid (DNA). Expression of eukaryotic DNA in prokaryotes is problematic for a variety of reasons including the requirement that the passenger DNA gene be free of intervening sequences and the possible instability of the messenger ribonucleic acid (mRNA) transcript or protein product in the prokaryote. In addition, some eukaryotic proteins require modification (for example, glycosylation) for normal biological activity, and appropriate modification of eukaryotic proteins is unlikely in a prokaryotic cell environment. Thus the genetic engineer may find it advantageous to use a eukaryotic host-vector system as the technological sophistication of genetic engineering increases.

The development of eukaryotic host-vector systems is under way. The primary eukaryotic viruses being pursued as possible transducing vectors in mammalian cells (SV40 and polyoma) and plant cells (cauliflower mosaic virus) are limited in the size of the passenger (exogenous) DNA that can be packaged in the virus particle, owing to the morphology of the nucleocapsid structure. Because the size of a eukaryotic gene is significantly increased by the presence of intervening sequences, the need for vectors that can accommodate more passenger DNA is accentuated. Furthermore, as the genetic engineering technology becomes more sophisticated, there will be an increased interest in inserting more than one passenger gene into a host cell to achieve coordinated expression and possibly obtain coordinated activity of the passenger gene products. Although "selectable genes" (for example,

Published with the approval of the director of the Idaho Agricultural Experiment Station as Research Paper no. 80519.

thymidine kinase or dihydrofolate reductase genes) have been suggested as possible means of introducing larger segments of exogenous DNA into eukaryotic cells, such systems lack several advantages of viral vector systems.

Using an ideal viral vector, it should be possible to introduce a large segment of passenger DNA into cells at high frequency; convert virtually all the cellular protein biosynthesis to the high-level expression of the foreign gene; and retain the ability to rapidly remove the passenger genes, modify them, and reinsert them as necessary. A virus that appears to be ideally suited as a vector for the propagation and high-level expression of many passenger genes in a higher eukaryotic environment is the baculovirus *Autographa californica* nuclear polyhedrosis virus (AcNPV).

In this chapter the properties of this interesting virus are discussed with particular emphasis on those features that make AcNPV an excellent candidate as a vector for transducing exogenous genes into a higher eukaryotic environment. Among these features are the potential for (1) inserting large segments of passenger genes into the viral DNA genome while retaining the ability to package the vector and passenger DNA in rod-shaped nucleocapsid structures, (2) obtaining high-level expression of inserted passenger DNA, (3) expressing invertebrate genes as well as possibly plant and animal genes, and (4) mass producing host cells and virus. In addition, there are significant safety features associated with nuclear polyhedrosis viruses (NPVs) as vectors for recombinant DNA work. More general information on the subjects of baculovirus structure and process of infection is available in several recent reviews (Carstens 1980; Harrap and Payne 1979; Summers 1978; Tinsley and Harrap 1978).

THE MORPHOLOGY OF AcNPV

AcNPV is a member of the baculovirus group within the family Baculoviridae (Matthews 1979). The term *baculovirus* derives from the rod-shaped nucleocapsid structures characteristic of this family. Nucleocapsids of AcNPV are 32 nm in diameter and 220 nm in length (Ramoska and Hink 1974). From a genetic engineering perspective, a cylindrical nucleocapsid structure suggests that essentially no restrictions are placed on the amount of passenger DNA that might be packaged along with the viral DNA in the capsid. Indeed, aberrant extended capsid structures have been observed during the replication of AcNPV and its closely related variants in cell culture (MacKinnon et al. 1974; Hirumi, Hirumi, and McIntosh 1975), indicating that the protein capsid structure does not have an inherent limit with regard to length and that the length of the mature nucleocapsid is probably dictated by the length of the DNA packaged. Fig. 14.1 shows some of

FIG. 14.1. Long, empty, capsidlike structures produced during aberrant replication of AcNPV L-1 in the nuclei of *Spodoptera frugiperda* cells at 33°C (H. H.
Lee and L. K. Miller). The cylindrical nature of AcNPV nucleocapsids and the observed extension of capsid length during aberrant replication of AcNPV suggest that
the capsid morphology will not be a limiting factor with regard to the quantity of
passenger DNA that may be packaged with the viral vector.

the long aberrant capsid structures observed during replication of AcNPV at the unusually high temperature of 33°C.

AcNPV assumes two mature forms that are referred to as the nonoccluded virus (NOV) form and the occluded virus (OV) or polyhedral inclusion body (PIB) form. Both of these forms have distinct biological roles in the baculovirus infection process (*vide infra*). In both the NOV and OV forms, the nucleocapsids are enveloped by a unit membrane, but the precise nature of the envelopes differs (Summers and Volkman 1976; Volkman, Summers, and Hsieh 1976). The nucleocapsids of NOVs acquire an envelope by budding through cellular membranes, most notably the plasma membrane, whereas the nucleocapsids of OVs acquire a membrane envelope within the nucleus (Adams, Goodwin, and Wilcox 1977; Hirumi, Hirumi, and McIntosh 1975; MacKinnon et al. 1974). In the case of AcNPV, multiple nucleocapsids are usually enclosed within a single envelope when envelopment occurs in the nucleus. Enveloped nucleocapsids within the cell nucleus may then be embedded in a crystalline protein matrix. The resulting protein crystals, containing many enveloped nucleocapsids, are known as occlusion bodies, PIBs, or, simply, polyhedra. A scanning electron microscope photograph of AcNPV occlusion bodies is presented in Fig. 14.2. The occlusion bodies of AcNPV are approximately 4 μm in diameter, allowing visualization with the light microscope.

In Fig. 14.3 a cross section through an AcNPV occlusion body in the process of formation is presented to illustrate the presence of multiple virions (enveloped nucleocapsids) embedded in the crystalline protein matrix of the occlusion body. Each virion consists of one or more nucleocapsids enveloped in a single membrane. This morphological feature (more than one nucleocapsid per envelope) is characeristic of a type of NPV frequently abbreviated as MNPV or NPV-MEV (multiple embedded virus) to distinguish it from NPVs in which a single nucleocapsid (the SNPV variety) is individually enveloped. From a genetic standpoint it is important that the presence of multiple nucleocapsids per envelope is characteristic of only the occluded form of an MNPV and not the NOV form since it is possible to clone and genetically manipulate the virus with ease using the NOV form (Lee and Miller 1978).

THE PROCESS OF AcNPV INFECTION

Understanding the roles of NOVs and OVs in the AcNPV infection process relates to why the genes controlling OV formation may be disposable and why the genetic engineer may be able to take advantage of the elimination or replacement of these genes to provide a safe vector system capable of high-level passenger gene expression. OVs are responsible for the horizontal

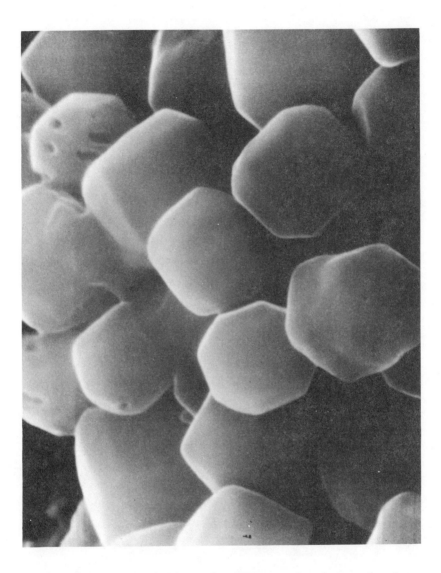

FIG. 14.2. The occluded form of AcNPV viewed in the scanning electron microscope (P. Treuer and L. K. Miller). Occlusion bodies of AcNPV are approximately 4 μm in diameter and contain many virions embedded within the crystalline protein matrix. The viral genes controlling occlusion may be replaced by the genetic engineer to achieve high-level expression of passenger genes while increasing the safety margin by making the virus defective in horizontal transmission among host organisms.

FIG. 14.3. A cross section of an AcNPV occlusion body in the process of formation in the nucleus of a *Spodoptera frugiperda* cell viewed by transmission electron microscopy (H. H. Lee and L. K. Miller). Each occlusion body of AcNPV contains several hundred virions. Each virion of AcNPV consists of one or more nucleocapsids enclosed within a single envelope. The envelopes of occluded virions differ biochemically and morphologically from envelopes of nonoccluded virions.

spread of AcNPV from organism to organism, whereas NOVs are responsible for the spread of infection within a single organism or in a cell culture. The host acquires the primary infection by ingesting material (for example, food) contaminated with OVs. The crystalline protein matrix of the OV apparently protects the virions within the OV from environmental factors as well as protecting the virions in the required passage through the foregut of the host to the midgut where the alkaline pH and proteolytic enzymes aid in the disruption of the crystalline protein matrix, resulting in the release of the virions from the occlusion body (Fig. 14.4A). Disruption of the OV may be achieved in vitro by incubation in alkaline solutions at pH values exceeding 10.5.

The virions released from the occlusion body attach to the epithelial cells lining the midgut to initiate the primary infection. Fig. 14.4 outlines the steps involved in the AcNPV infection process and is derived from a compilation of literature dealing with AcNPV and its closely related variants *Trichoplusia ni* NPV (TnMNPV) and *Rachiplusia ou* NPV (RoMNPV) (Adams, Goodwin, and Wilcox 1977; Falcon and Hess 1977; Hirumi, Hirumi, and McIntosh 1975; Kawanishi et al. 1972; MacKinnon et al. 1974). Following fusion of the viral envelope and the midgut cell membrane (Fig. 14.4B), the nucleocapsids enter the cytoplasm (Fig. 14.4C) and travel to the nuclear membrane (Fig. 14.4D). Nucleocapsids of AcNPV apparently enter the nucleus and uncoat within the nucleus (Adams, Goodwin, and Wilcox 1977; Hirumi, Hirumi, and McIntosh 1975; Kawanishi et al. 1972) in a similar fashion to *Heliothis zea* NPV (Granados 1978). The precise mechanism of entry into the nucleus is not yet known but may occur by both entry through a nuclear pore or budding through the nuclear membrane. The viral DNA (Fig. 14.4E) normally replicates in the nucleus with the resultant production of progeny nucleocapsids (Fig. 14.4F). Accompanying this replication process is the dispersal of the cell chromatin, disappearance of the nucleolus, and the appearance of the virogenic stroma (Fig. 14.4F), the site of viral DNA replication, and nucleocapsid formation.

In the midgut cells most of the progeny nucleocapsids appear to leave the cell by budding through cell membranes. Nucleocapsids budding or blebbing through the nuclear membrane (Fig. 14.4G) and vesicles containing nucleocapsids have been observed in the cytoplasm (Fig. 14.4H), but the vesicles may be destroyed before reaching the plasma membrane. Most nucleocapsids observed leaving the cell do so by budding through the plasma membrane (Fig. 14.4I). The nucleocapsids usually bud from the plasma membrane of the cell individually. The singly enveloped NOV enter the circulatory system of the host and proceed to infect other tissues, setting up a systemic infection in the organism.

Plasma membrane budded (PMB) NOVs released into the circulatory

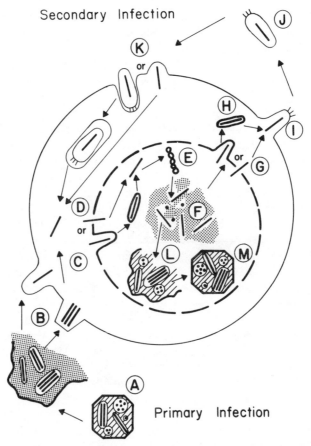

FIG. 14.4. An outline of the process of AcNPV infection. The primary infection of an organism begins in the midgut where occlusion bodies (**A**) are disrupted by alkaline pH and proteolytic enzymes, thereby releasing multiple virions into the gut lumen (**B**). The envelopes of the virions fuse with the plasma membrane of the midgut epithelial cells, and the nucleocapsids travel through the cytoplasm (**C**) to the nucleus. Following entry (**D**) and uncoating in the nucleus, the double-stranded, circular DNA (**E**) replicates and progeny nucleocapsids are formed (**F**). Nucleocapsids formed early in the replication process can leave the nucleus (**G** or **H**) and subsequently bud through the plasma membrane (**I**) into the circulatory system (or cell culture media), acquiring peplomeric structures in the process. The PMB-NOV (**J**) may then initiate secondary infection following entry into cells by membrane fusion or viropexis (**K**). Nucleocapsids formed in cell nuclei during secondary infection have the additional option of being enveloped (**L**) in the nucleus and occluded there (**M**). The budding of nucleocapsids from the plasma membrane precedes the occlusion process by approximately 4 hours. Occlusion bodies are observed in cell cultures by 14 to 18 hours postinfection (see text for further detail and relevant references).

system or into cell culture medium are different in many respects from the virions released from alkaline-disrupted occlusion bodies (Summers and Volkman 1976). The membrane of an NOV is loosely arranged around the single nucleocapsid, and peplomeric structures are observed at the anterior end (Fig. 14.4J) (Adams, Goodwin, and Wilcox 1977; Summers and Volkman 1976). NOVs are observed to enter cells in culture by either membrane fusion or viropexis (Fig. 14.4K), and they proceed to establish a full infection producing both NOV and OV in susceptible host cells such as *Spodoptera frugiperda* (IPLB-SF-21) or *Trichoplusia ni* (TN-368). A similar process presumably also occurs in tissue involved in the secondary infection mediated by PMB-NOV in host organisms.

The envelopes of occluded virions have their origin in the nucleus and do not appear to have the peplomeric structures observed at the anterior end of PMB-NOVs. The nuclear-formed virions also lack at least one antigenic determinant possessed by PMB-NOV (Volkman, Summers, and Hsieh 1976). Nucleocapsids that are enveloped in the nucleus (Fig. 14.4L) may be occluded in polyhedral inclusion bodies formed within the nucleus (Fig. 14.4M). Occlusion of virions in the midgut cells is greatly reduced compared with cells in secondarily infected tissues (Harrap and Robertson 1968); this tissue-specific effect would be expected to facilitate systemic infection since more nucleocapsids would be available for extracellular NOV production. In cell culture NOV release usually precedes occlusion temporally (*vide infra*). In the case of AcNPV in nature, the infection is so pervasive that as much as 10 percent of the weight of a dead infected host organism may be accounted for as occluded virus. OVs are released upon the death and disintegration of the host organism.

OVs are not infectious in cell culture unless virions are released from the occlusion body by artificial alkaline disruption. Even under these conditions, the relative infectivity of alkali-released virions (plaque forming unit [pfu]/particle ratio) is very low compared with PMB-NOVs (Volkman and Summers 1977). Consequently, OVs formed in cell culture are noninfectious in cell cultures, and NOVs are responsible for the spread of infection from cell to cell. The NOVs are relatively noninfectious to the host organism unless injected into the circulatory system.

From a genetic engineering standpoint, the genes controlling occlusion are unnecessary, and the elimination or replacement of these genes would be desirable for two reasons (1) the resulting virus would be defective in horizontal transmission at the organism level but nondefective in NOV replication in cell culture thus providing a considerable safety. margin in terms of virus escape and survival in nature and (2) substitution of passenger DNA into the region controlling occlusion (with appropriate attention to promoter regions, et cetera) could provide a high level of expression of passenger genes at late times in the infection cycle (*vide infra*).

POLYHEDRIN

The crystalline protein constituting the matrix of the occlusion body of AcNPV is predominantly a single protein that is approximately 30,000 daltons and is known as polyhedrin (Summers and Smith 1978). Roughly 95 percent of the weight of an occlusion body consists of polyhedrin. Since as much as 10 percent of the body weight of a dead host is occluded virus, polyhedrin is synthesized in truly enormous quantities. Another way of viewing this remarkable level of synthesis of a single protein is to consider that cells infected with a wild type strain of AcNPV produce an average of over 60 occlusion bodies per cell, each occlusion body being approximately 4 μm in diameter. There is no indication of an amplification of the poly-hedrin gene in the process of replication, as determined by restriction endo-nuclease analysis and Southern blotting of the DNA of infected cells (Tjia, Carstens, and Doerfler 1979). It is therefore likely that the promoter for polyhedrin mRNA synthesis is exceptionally strong. The genetic engineer may find it advantageous to utilize this promoter for high levels of passen-ger gene expression.

The amino acid sequence of one NPV polyhedrin is known and the N-terminal sequence of several other NPV polyhedrins is also available (Rohrman et al. 1979); Serebryani et al. 1977; G. F. Rohrman 1981: per-sonal communication). A common feature of the N-termini of NPV poly-hedrins is a high proportion of tyrosine residues, which may account for the alkaline dissolution properties of the occlusion bodies. The secondary struc-ture of the N-terminus may consist of four β turns with small β sheets on either side leading into an α helix.

The polyhedrin gene is encoded by the AcNPV genome as indicated by the hybridization selection of viral-specific mRNAs followed by in vitro translation and precipitation of a 30,000-dalton protein product, using anti-polyhedrin antibody (Van Der Beek, Saaÿer-Riep, and Vlak 1980). The position of the gene for polyhedrin has been located within a 20 percent region of the AcNPV physical map by intertypic recombination experi-ments (Summers et al. 1980) (vide infra). A further localization and charac-terization of this gene will be aided by cloning of the complementary DNA (cDNA) synthesized from the mRNA coding for polyhedrin and comparing the sequence of this DNA with the equivalent DNA regions of the viral DNA genome.

THE DNA GENOME OF AcNPV

The genome of AcNPV is a double-stranded, circular, convalently closed DNA molecule of approximately 82 to 88 million daltons (Miller and Dawes

1979; Smith and Summers 1979; Summers and Anderson 1973). Physical maps of the sites recognized by *Bam*HI, *Eco*RI, *Xma*I (*Sma*I), *Hind*III, *Sac*I (*Sst*I), *Xho*I, and *Kpn*I are available (Miller and Dawes 1979; Smith and Summers 1979). A combined physical map is presented in Fig. 14.5.

The DNA genome of AcNPV is approximately 41.8 ± 0.5 percent guanine plus cytosine as estimated by buoyant density analysis in CsCl (1.7010 ± 0.005 g/ml) to 43.6 ± 0.3 percent guanine plus cytosine as estimated from the Tm ($71.8 \pm 0.1°$C in $0.1 \times$ SSC) (Vlak and Odink 1979). There is no indication of repeated sequence DNA in the AcNPV genome from experiments involving hybridization of [^{32}P]-labeled restriction endonuclease fragments of AcNPV to Southern blots of fractionated restriction endonuclease fragments of AcNPV DNA (Miller and Dawes 1979; Summers et al. 1980). However, these experiments do not rule out the existence of short, tandemly repeated sequences. An electron microscopic estimation of the molecular weight of AcNPV DNA (Tjia, Carstens, and Doerfler 1979) is unaccountably high (92 million daltons) relative to the molecular weight estimates (82 and 88 million daltons) by restriction endonuclease fragment mapping (Miller and Dawes 1979; Smith and Summers 1979). There are no 5'-C^{m}CGG-3' sequences in AcNPV DNA, suggesting that the DNA is not extensively methylated (Tjia, Carstens, and Doerfler 1979).

A number of genotypic variants of AcNPV have been observed (Lee and Miller 1978; Miller and Dawes 1978; Smith and Summers 1978; Tjia, Carstens, and Doerfler, 1979), and restriction endonuclease site variations for several of the naturally occurring variants have been physically mapped (Miller and Dawes 1979; Summers et al. 1980; Smith and Summers 1979). The variant *Rachiplusia ou* NPV (RoMNPV) shares approximately 97 percent sequence homology with AcNPV throughout its DNA genome (Jewell and Miller 1980; Summers et al. 1980). Differences between several of the structural proteins of these two viruses have been noted (Summers and Smith 1978). Smith and Summers (1980) extensively mapped the restriction sites of the RoMNPV variant, and Summers et al. (1980) utilized the observed protein and DNA differences to physically map the polyhedrin gene within a 20 percent region of the AcNPV genome using the intertypic mixing technique developed for adenovirus recombinants (Williams et al. 1975). The polyhedrin gene lies within the 76 to 96 percent region of the physical map shown in Fig. 14.5. Owing to several difficulties, including the low proportion of AcNPV/RoMNPV recombinants, the frequency of recombinants exhibiting a common crossover site, and the sparsity of usable genotypic markers (Summers et al. 1980), a more precise mapping of polyhedrin was not possible. Several other structural proteins have been located within broader regions of the DNA genome (Summers et al. 1980).

The AcNPV DNA genome is infectious in cell cultures using modifications of the CaCl$_2$ technique employed in mammalian cells (Burand, Sum-

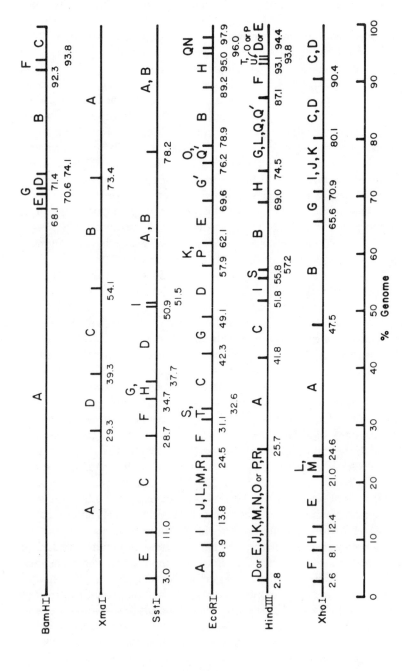

FIG. 14.5. A physical map of AcNPV L-1 compiled from data of Miller and Dawes (1979) and Smith and Summers (1979).

214

mers, and Smith 1980; Carstens, Tjia, and Doerfler 1979; Potter and Miller 1980b) or a DEAE-dextran technique (Potter and Miller 1980b). Specific infectivities of AcNPV DNA using the $CaCl_2$ techniques were as high as 6.1×10^4 pfu/μg DNA. Specific infectivities using the DEAE-dextran technique were generally lower than the specific infectivities achieved with the $CaCl_2$ techniques.

The ability to transfect cell cultures with relatively high efficiency (considering the large size of the AcNPV DNA genome) is important with respect to genetic engineering for at least two reasons: (1) it should be possible to ligate passenger DNA to the viral DNA, insert the recombinant DNA into cultured cells by a transfection procedure, and obtain virus containing the additional DNA sequences and (2) the DNA of AcNPV or a recombinant can be manipulated in vitro and reinserted into cells with relative ease. It may also be noted that the large size of the AcNPV DNA genome is advantageous with regard to the introduction of large passenger DNA segments into cells. Selecting for infectivity involves selection for the uptake of 80 million daltons of DNA; an additional 20 million or so daltons of passenger DNA should have minimal effect on overall specific infectivity.

THE GENETICS OF AcNPV

The ability to clone AcNPV was first demonstrated by the isolation of genotypic variants of AcNPV from a genotypically heterogenous stock (Lee and Miller 1978). Important to this work was the use of NOV, with a single nucleocapsid per envelope, as the inoculum in cell culture. A number of simple and excellent plaque assay techniques are available for genetic work (Brown and Faulkner 1978; Lee and Miller 1978; Wood 1977).

Temperature-sensitive (ts) mutants of AcNPV have been isolated and partially characterized (Brown, Crawford, and Faulkner 1979; Lee and Miller 1979). Several different phenotypes have been observed including lack of OV production (NOV$^+$,OV$^-$), defective NOV production (NOV$^-$,OV$^+$), defective NOV and OV production (NOV$^-$,OV$^-$), and defective DNA replication (DNA$^-$). The existence of NOV$^+$,OV$^-$ mutants indicates that at least certain genes controlling OV formation are nonessential and may be deleted for genetic engineering purposes. Indeed, the number of mutants with this phenotype and the corresponding number of complementation groups suggest that a significant proportion of the AcNPV genome may be involved in the process of occlusion (Lee and Miller 1979). Electron microscopic analyses of the morphogenesis of AcNPV and TnNPV occlusion body formation (Chung, Brown, and Faulkner 1980; Summers and Arnott 1969) suggest a complex assembly mechanism that may involve numerous genes. The regions of DNA exclusively involved in the occlusion process will probably best be determined by the isolation of deletion mutants defective in OV formation.

Mapping the ts mutants with respect to the physical map by a marker rescue technique (Potter and Miller 1980a) should provide insight into the genetic organization of these viruses. The marker rescue method for mapping ts mutants of AcNPV is based on the recombination in vivo of a mutant DNA genome with a restriction fragment of wild type AcNPV DNA. Potter and Miller (1980a) demonstrated the ability to map ts mutants of AcNPV (having either an NOV⁻ or OV⁻ phenotype) using the marker rescue technique. Thus far, seven ts mutants of AcNPV have been mapped, and the precision of mapping that may be achieved with this technique is localization to a region less than 1 percent of the genome (Miller, manuscript in preparation). Establishing a genetic map of AcNPV based on the physical restriction map of restriction endonuclease sites is essential for an understanding of the gene organization of the virus and the eventual sophisticated manipulation of the DNA genome by the genetic engineer.

THE TEMPORAL CONTROL OF THE AcNPV INFECTIOUS PROCESS

Studies of AcNPV growth in cloned cell lines of *Trichoplusia ni* (Tn-368) demonstrate a definite sequence of events in the production of NOV and OV forms (Volkman, Summers, and Hsieh 1976). Release of NOV into culture media begins approximately 10 hours postinfection and coincides with observable cytopathic effects (CPE). OV production begins approximately 14 hours postinfection. As OV production increases, the rate of NOV production decreases. Significant levels of cytolysis are not observed until 48 hours postinfection. Thus, there is approximately a 4-hour delay between the onsets of NOV and OV production. A long period of time follows the onset of OV formation in which the cell remains metabolically active.

The actual synthesis of the proteins involved in NOV and OV formation was shown by an immunoperoxidase technique to be temporally controlled (Summers, Volkman, and Hsieh 1978). NOV antigens were detected at 6 to 8 hours postinfection (approximately 2 hours before the release of NOV into the culture media), whereas polyhedrin antigens were detected at 12 hours postinfection (approximately 2 hours before the observable onset of OV formation). Thus, there is a temporal sequence in the synthesis of structural virus proteins, with polyhedrin synthesis beginning approximately 2 hours after the onset of NOV production.

Several studies on the time course of protein synthesis in AcNPV-infected cell cultures have indicated a sequential appearance of viral-induced polypeptides as determined from [^{35}S]-methionine pulse labeling experiments (Carstens, Tjia, and Doerfler 1979; Dobos and Cochran 1980; Wood 1980). Although there is general agreement that early, middle, and late periods of protein synthesis can be observed, many of the specifics such as how many, what size, and at what time the proteins are synthesized will need further clarification. The first new proteins synthesized upon AcNPV infection are observed as early as 2 to 3 hours postinfection (Carstens, Tjia, and Doerfler 1979; Wood 1980). DNA synthesis begins approximately 5 hours following infection at a high multiplicity (Tjia, Carstens, and Doerfler 1979). Other proteins appear at 10 hours postinfection, and some are delayed until 16 to 22 hours postinfection.

In two studies polyhedrin synthesis was first observed 16 to 22 hours postinfection and no posttranslational cleavage was observed (Dobos and Cochran 1980; Wood 1980). Differences in cell lines, multiplicities of infection, and sensitivities of detection methods may account for the time differences observed between the [^{35}S]-methionine labeling and immunoperoxidase studies. The sequential protein synthesis in AcNPV-infected cells is independent of DNA replication as indicated by the observation that cytosine arabinoside did not inhibit the synthesis of viral proteins for the formation of OVs (Dobos and Cochran 1980).

The sequence of NOV production, polyhedrin synthesis, and OV formation may be important for the genetic engineer interested in producing large quantities of passenger gene products. As noted earlier, polyhedrin is made in enormous quantities in infected cells, and thus the genetic engineer may find it advantageous to replace the polyhedrin gene with passenger DNA, utilizing the polyhedrin promoter for expression. An additional advantage of placing passenger DNA in this position is that polyhedrin synthesis occurs late in the infection cycle, subsequent to the onset of NOV release and more than 24 hours prior to cytolysis. This is an ideal situation since infectious virus production would not be impaired by any active passenger gene product and a long period of cell viability suggests continued metabolic functioning for passenger protein synthesis and enzymatic functioning if desired.

SAFETY OF NUCLEAR POLYHEDROSIS VIRUSES

Presently there are three NPVs registered by the U.S. Environmental Protection Agency (EPA) as pesticides. One of these, the *Heliothis zea* nuclear polyhedrosis virus (HzNPV), is commercially produced by Sandoz Corporation under the trade name Elcar and has been sprayed commercially since 1976 on millions of acres of land in the southern United States to control the cotton bollworm. Before registration as a pesticide by the EPA, HzNPV underwent a variety of tests over a 15-year period to determine its potential pathogenicity for nontarget organisms (Ignoffo 1973,1975; McIntosh and Maramorosch 1973). The tests have included various routes of administration including *per os*, inhalation, dermal application, intradermal, intramuscular, intracerebral, intravenous, intraperitoneal, and subcutaneous. Test animals have included mouse, rat, guinea pig, rabbit, dog, monkey, man, quail, chicken, sparrow, mallard, and a variety of fish. By 1974 an estimated 1 million acres of crops including corn, sorghum, tomatoes, radishes, cabbage, beans, and tobacco had been sprayed with NPVs with no observed phytotoxicity or pathogenicity. To date this estimate of acreage must be increased many fold due to commercial spraying in North America as well as test spraying of a variety of NPVs in other continents including Europe, Africa, South America, Australia, and Asia.

AcNPV is a candidate for EPA registration and has undergone a series of safety tests relevant to registration as a microbial pesticide. Although this type of safety testing has been designed with pesticide use rather than vector use in mind, the testing is certainly supportive of the relative safety of NPVs as previously noted (Miller 1977). Indeed, any virus that is approved by the EPA for massive dissemination must have a few safety features relevant to use as a vector for recombinant DNA research. The two most important features of baculoviruses in terms of their safety as pesticides or as recombinant DNA vectors are: (a) baculoviruses are known to infect only invertebrates (there are no known vertebrate or plant baculoviruses); and (b) baculoviruses exhibit a relatively high degree of host specificity, each virus usually confined to a host genus or several genera. The AcNPV has a broader host range than many NPVs, but it is nevertheless limited to a few families in the order Lepidoptera; essentially all hosts of AcNPV are considered pest species. The one report of AcNPV replication in a dipteran (mosquito) cell culture (Sherman and McIntosh 1979) must be repeated with appropriate controls. The extended labeling period (4 days) with [^{3}H]-thymidine would result in distribution of label to molecules other than thymidine. Furthermore, since the encapsidation of AcNPV is independent of DNA replication as demonstrated in a lepidopteran cell line, [^{3}H]-thymidine incorporation might only represent DNA repair rather than DNA replication (Dobos and Cochran 1980). It would be truly exciting if AcNPV could replicate in dipteran cultures, particularly *Drosophila melanogaster* cell lines, but this possibility seems to be remote.

An additional safety consideration of relevance to the use of AcNPV as a vector for recombinant DNA is that occlusion is essential for the natural horizontal transmission of AcNPV at the organism level, and if a virus is used that is defective (preferably by DNA deletion) in occlusion, the spread of the virus in nature will be prevented.

MASS PRODUCTION OF AcNPV

Since AcNPV is considered to have excellent potential as a biological pesticide, research has been aimed at the mass production of this virus for pesticide use. Commercial quantities of AcNPV are currently produced by Sandoz Corporation in mass-reared host larvae. Research was also undertaken to develop means for the mass production of AcNPV in cell culture (Vaughn and Goodwin 1977). If large quantities of passenger DNA gene products are desired, such cell culture systems are available.

POSSIBLE USES OF AN AcNPV TRANSDUCING VECTOR

An invertebrate host-vector system has the potential of being extremely versatile with possible uses in plant and vertebrate as well as in invertebrate research or applied technology. Invertebrates, particularly insects, may play a central or pivotal role in natural genetic exchange among plants and vertebrates. Insects are known to act as vectors of both plant and animal viruses indicating that, at the biochemical level, insects share many common features with both plants and animals. This suggests that there is a good probability that plant or animal DNA will be properly expressed, appropriately modified, and enzymatically active in the invertebrate environment. It is less probable that the passenger gene products will interact intimately with host proteins or become involved in ongoing catalytic processes of the host cell that require extensive steric interactions.

Heterologous systems (for example, mammalian passenger DNA in an invertebrate host) should have advantages not found in homologous systems (for example, mammalian passenger DNA in a mammalian host). For example, the gene products of a given passenger DNA may disturb the metabolism (for example, shut off protein synthesis) in the homologous host but not in the heterologous host. Another advantage of using heterologous systems is that the exogenous gene product may be more easily identified (biochemically, immunologically, et cetera) in the heterologous host.

Homologous systems are most useful in research analyzing the expression and regulation of specific genes in their homologous environment. Since insects provide a variety of excellent model systems for studying the molecular basis of development and behavior, an invertebrate host-vector system will be valuable in the homologous context as well. The AcNPV vector may be particularly applicable in studies with lepidoptera such as *Bombyx mori* (a host of AcNPV but nonpermissive for OV formation) and *Antheraea polyphemus* (Kafatos et al. 1977).

Two additional features of an AcNPV vector are the large size of passenger DNA that may be inserted in the vector and the potential of high levels of passenger gene expression. The need for vectors with these capabilities will increase as genetic engineering technology progresses and the interest in pursuing more complex genetic transfers increases for basic or applied reasons. Of particular interest may be production in cell culture of natural products from eukaryotes requiring several active proteins for synthesis. An example of the latter would be the production of pheromones for commercial use as biological pesticides.

PROBLEMS IN USING AcNPV AS A VECTOR

Probably the only major problem in developing AcNPV as a vector is that unlike the mammalian viruses such as the papovaviruses and adenoviruses that are being developed as vectors for recombinant DNA research, AcNPV is not known to cause tumors in mammalian cells. As a result very little money and very few people are devoted to studying the molecular biology of AcNPV in comparison with the tumor virus fields. Until further information is available regarding the basic gene organization of AcNPV and the factors controlling occlusion, the use of this virus vector system will be delayed. Hopefully, this brief introduction to the possibilities offered by this fascinatingly complex virus will serve to stimulate interest in the pursuit of research on AcNPV.

The large size of AcNPV provides certain advantages with regard to vector use (*vide supra*), but it is also a disadvantage since the vast genome size implies a complex infection pathway. Preliminary genetic analysis, however, indicates that much of the AcNPV genome may control aspects of occlusion, a potentially dispensable process for vector use. Without the occlusion process the basic mechanism of AcNPV may be relatively simple, and thus, the work required in understanding the basic gene organization before employment as a vector may be greatly reduced. Nevertheless, understanding the control of occlusion will be important if the polyhedrin promoter is utilized for the high-level expression of passenger genes.

SUMMARY

Many features of the baculovirus AcNPV indicate the potential usefulness of this virus as a vector for transducing exogenous genes into an invertebrate host. These features include a rod-shaped capsid that may accommodate 20 megadaltons or more of passenger DNA and a sequential expression of viral genes resulting in the early production of infections virus followed by a long period of extensive virus-directed protein synthesis.

The basic framework for developing AcNPV as a vector has been established. Thus the physical map of the DNA genome is available, and the DNA genome is infectious in cell culture using a $CaCl_2$ transfection procedure. Some mutants, variants, and recombinants of AcNPV have been isolated and the genetic alterations mapped relative to the physical map. Information is becoming available on the proteins produced upon infection. A key position for inserting passenger DNA has been identified.

Additional attractive features regarding the use of AcNPV as a vector include the relative safety of the virus and the possible use of large cell culture systems for producing gene products on a mass scale. Plant and vertebrate as well as invertebrate genes may be advantageously propagated and expressed in an invertebrate background.

REFERENCES

Adams, J. R.; Goodwin, R. H.; and Wilcox, T. A. 1977. Electron microscopic investigations on invasion and replication of insect baculoviruses *in vivo* and *in vitro*. *Biologie Cellulaire* 28:261–68.

Brown, M.; Crawford, A. M.; and Faulkner, P. 1979. Genetic analysis of a baculovirus, *Autographa californica* nuclear polyhedrosis virus. *J. Virol.* 31:190–98.

Brown, M., and Faulkner, P. 1978. Plaque assay of nuclear polyhedrosis viruses in cell culture. *Appl. Environ. Microbiol.* 36:31–35.

Burand, J. P.; Summers, M. D.; and Smith, G. E. 1980. Transfection with baculovirus DNA. *Virology* 101:286–90.

Carstens, E. 1980. Baculoviruses—friend of man, foe of insects? *Trends in Biochem. Sci.* 52:107–10.

Carstens, E. B.; Tjia, S. T.; and Doerfler, W. 1979. Infection of *Spodoptera frugiperda* cells with *Autographa californica* nuclear polyhedrosis virus. *Virology* 99:386–98.

———. 1980. Infectious DNA from *Autographa californica* nuclear polyhedrosis virus. *Virology* 101:311–14.

Chung, K. L.; Brown, M.; and Faulkner, P. 1980. Studies on the morphogenesis of polyhedral inclusion bodies of a baculovirus, *Autographa californica* NPV. *J. Gen. Virol.* 46:335–47.

Dobos, P., and Cochran, M. A. 1980. Protein synthesis in cells infected by *Autographa californica* nuclear polyhedrosis virus (Ac-NPV): the effect of cytosine arabinoside. *Virology* 103:446–64.

Falcon, L. A., and Hess, R. T. 1977. Electron microscope study on the replication of *Autographa* nuclear polyhedrosis virus and *Spodoptera* nuclear polyhedrosis virus in *Spodoptera exigua*. *J. Invert. Pathol.* 29:36–43.

Granados, R. R. 1978. Early events in the infection of *Heliothis zea* midgut cells by a baculovirus. *Virology* 90:170–74.

Harrap, K. A., and Payne, C. C. 1979. The structural properties and identification of insect viruses. In *Advances in Virus Research*, ed. M. A. Lauffer, F. B.

Bang, K. Maramorosch, and K. M. Smith, vol. 25, pp. 273–355. New York: Academic Press.

Harrap, K. A., and Robertson, J. S. 1968. A possible infection pathway in the development of a nuclear polyhedrosis virus. *J. Gen. Virol.* 3:221–25.

Hirumi, H.; Hirumi, K.; and McIntosh, A. H. 1975. Morphogenesis of a nuclear polyhedrosis virus of the alfalfa looper in a continuous cabbage looper cell line. *Ann. N.Y. Acad. Sci.* 266:302–26.

Ignoffo, C. M. 1973. Effects of entomopathogens on vertebrates. *Ann. N.Y. Acad. Sci.* 217:141–72.

——— . 1975. Evaluation of *in vivo* specificity of insect viruses. In *Baculoviruses for insect pest control: safety considerations*, ed. M. Summers, R. Engler, L. A. Falcon, and P. Vail, pp. 52–60. Washington, D.C.: American Society for Microbiology.

Jewell, J. E., and Miller, L. K. 1980. DNA sequence homology relationships among six lepidopteran nuclear polyhedrosis viruses. *J. Gen. Virol.* 48: 161–75.

Kafatos, F. C.; Maniatis, T.; Efstratiadis, A.; Kee, S. G.; Regier, J. C.; and Nadel, M. 1977. The moth chorion as a system for studying the structure of developmentally regulated gene sets. In *The organization and expression of the eukaryotic gene*, ed. E. M. Bradbury and K. Javaherian, pp. 393–420. London: Academic Press.

Kawanishi, C. Y.; Summers, M. D.; Stoltz, D. B.; and Arnott, H. J. 1972. Entry of an insect virus *in vivo* by fusion of viral envelope and microvillus membrane. *J. Invert. Pathol.* 20:104–8.

Lee, H. H., and Miller, L. K. 1978. Isolation of genotypic variants of *Autographa californica* nuclear polyhedrosis virus. *J. Virol.* 27:754–67.

——— . 1979. Isolation, complementation and initial characterization of temperature-sensitive mutants of the baculovirus *Autographa californica* nuclear polyhedrosis virus. *J. Virol.* 31:240–52.

McIntosh, A. H., and Maramorosch, K. 1973. Retention of insect virus infectivity in mammalian cell culture. *N.Y. Ent. Soc.* 81:175–82.

MacKinnon, E. A.; Henderson, J. F.; Stoltz, D. B.; and Faulkner, P. 1974. Morphogenesis of nuclear polyhedrosis virus under conditions of prolonged passage in vitro. *J. Ultrastruct. Res.* 49:419–35.

Matthews, R. E. F. 1979. Classification and nomenclature of viruses. *Intervirology* 12:134–296.

Miller, L. K. 1977. A safer model system for studying the effects of recombining animal virus DNA. In *Genetic manipulation as if affects the cancer problem*, ed. J. Schultz and Z. Drada, p. 271. New York: Academic Press. (Abstract)

Miller, L. K., and Dawes, K. P. 1978. Restriction endonuclease analysis to distinguish two closely related nuclear polyhedrosis viruses: *Autographa californica* MNPV and *Trichoplusia ni* MNPV. *J. Appl. Environ. Microbiol.* 35:1206–10.

——— . 1979. Physical map of the DNA genome of *Autographa californica* nuclear polyhedrosis virus. *J. Virol.* 29:1044–55.

Potter, K. N., and Miller, L. K. 1980a. Correlating genetic mutations of a baculovirus with the physical map of the DNA genome. In *Animal virus genetics*, ed. B. Fields, R. Jaenisch, and C. F. Fox, vol. 18, pp. 71–80. ICN-UCLA Symposia on Molecular and Cellular Biology. New York: Academic Press.

——. 1980b. Transfection of two invertebrate cell lines with DNA of *Autographa californica* nuclear polyhedrosis virus. *J. Invert. Pathol.* 36:431–32.

Ramoska, W. A., and Hink, W. F. 1974. Electron microscope examination of two plaque variants from a nuclear polyhedrosis virus of the alfalfa looper, *Autographa californica*. *J. Invert. Pathol.* 23:197–201.

Rohrman, G. F.; Bailey, G. F.; Brimhall, B.; Becker, R. R.; and Beaudreau, G. S. 1979. Tryptic peptide analysis and NH$_2$-terminal amino acid sequences of polyhedrins of two baculoviruses from *Orgyia pseudotsugata*. *Proc. Nat. Acad. Sci. (USA)* 76:4976–80.

Serebryani, S. B.; Levitina, T. L.; Kautsman, M. L.; Radavski, Y. L.; Gusak, N. M.; Ovander, M. N.; Sucharonko, N. V.; and Kozlov, E. A. 1977. The primary structure of the polyhedral protein of nuclear polyhedrosis virus of *Bombyx mori*. *J. Invert. Pathol.* 30:442–43.

Sherman, K. E., and McIntosh, A. H. 1979. Baculovirus replication in a mosquito (dipteran) cell line. *Infection and Immunity* 26:232–34.

Smith, G. E., and Summers, M. D. 1978. Analysis of baculovirus genomes with restriction endonucleases. *Virology* 89:517–27.

——. 1979. Restriction maps of five *Autographa californica* MNPV variants, *Trichoplusia ni* MNPV and *Galleria mellonella* MNPV DNAs with endonucleases SmaI, KpnI, BamHI, SacI, XhoI and EcoRI. *J. Virol.* 30:828–38.

——. 1980. Restriction map of *Rachiplusia ou* and *Rachiplusia ou-Autographa californica* baculovirus recombinants. *J. Virol.* 33:311–19.

Summers, M. D. 1978. Baculoviruses (Baculoviridae). In *The atlas of insect and plant viruses*, ed. K. Maramorosch, pp. 3–27. New York: Academic Press.

Summers, M. D., and Anderson, D. L. 1973. Characterization of nuclear polyhedrosis virus DNAs. *J. Virol.* 12:1336–46.

Summers, M. D., and Arnott, H. J. 1969. Ultrastructural studies on inclusion formation and virus occlusion in nuclear polyhedrosis and granulosis virus-infected cells of *Trichoplusia ni* (Hübner). *J. Ultrastruct. Res.* 28:462–80.

Summers, M. D., and Smith, G. E. 1978. Baculovirus structural polypeptides. *Virology* 84:390–402.

Summers, M. D.; Smith, G. E.; Knell, J. D.; and Burand, J. P. 1980. Physical maps of *Autographa californica* and *Rachiplusia ou* nuclear polyhedrosis virus recombinants. *J. Virol.* 34:693–703.

Summers, M. D., and Volkman, L. E. 1976. Comparison of biophysical and morphological properties of occluded and extracellular nonoccluded baculovirus from in vivo and in vitro host systems. *J. Virol.* 17:962–72.

Summers, M. D.; Volkman, L. E.; and Hsieh, C. H. 1978. Immunoperoxidase dectection of baculovirus antigens in insect cells. *J. Gen. Virol.* 40:545–57.

Tinsley, T. W., and Harrap, K. A. 1978. Viruses of invertebrates. In *Comprehensive virology*, ed. H. Fraenkel-Conrat and R. R. Wagner, vol. 12, pp. 1–101. New York: Plenum Press.

Tjia, S. T.; Carstens, E. B.; and Doerfler, W. 1979. Infection of *Spodoptera frugiperda* cells with *Autographa californica* nuclear polyhedrosis virus. *Virology* 99:399–409.

Van Der Beek, C. P.; Saaÿer-Riep, J. D.; and Vlak, J. M. 1980. On the origin of the polyhedral protein of *Autographa californica* nuclear polyhedrosis virus.

Virology 100:326–33.

Vaughn, J. L., and Goodwin, R. H. 1977. Large-scale culture of insect cells for virus production. In *Virology in agriculture, Beltsville symposia in agricultural research I*, ed. J. A. Romberger, pp. 109–16. Montclair, N.J.: Allanheld, Osmun.

Vlak, J. M., and Odink, K. G. 1979. Characterization of *Autographa californica* nuclear polyhedrosis deoxyribonucleic acid. *J. Gen. Virol.* 44:333–47.

Volkman, L. E., and Summers, M. D. 1977. *Autographa californica* nuclear polyhedrosis virus: comparative infectivity of the occluded, alkali-liberated, and nonoccluded forms. *J. Invert. Pathol.* 30:102–3.

Volkman, L. E.; Summers, M. D.; and Hsieh, C. H. 1976. Occluded and nonoccluded nuclear polyhedrosis virus grown in *Trichoplusia ni*: comparative neutralization, comparative infectivity and in vitro growth studies. *J. Virol.* 19:820–32.

Williams, J.; Grodzicker, T.; Sharp, P.; and Sambrook, K. J. 1975. Adenovirus recombination: physical mapping of crossover events. *Cell* 4:113–19.

Wood, H. A. 1977. An agar overlay plaque assay method for *Autographa californica* nuclear polyhedrosis virus. *J. Invert. Pathol.* 29:304–7.

——— . 1980. *Autographa californica* nuclear polyhedrosis virus-induced proteins in tissue culture. *Virology* 102:21–27.

15

POSSIBLE USES OF RECOMBINANT DNA FOR GENETIC MANIPULATIONS OF ENTOMOPATHOGENIC BACTERIA

Robert M. Faust
Charles F. Reichelderfer
Curtis B. Thorne

INTRODUCTION

Most increases in crop production have been achieved by scientists whose goals have been directly aimed at increasing plant yields. Within the past several decades, however, new knowledge and techniques have been developed in the sciences, especially in insect pathology, that hold promise for alternative methods of increasing crop production via crop protection. Although accurate figures on crop losses due to insect pests are somewhat difficult to obtain, these losses to world agricultural production have been estimated at more than $4 billion per year. Increasing problems with the use of chemical insecticides (that is, insect resistance, insecticidal residues, toxicity to nontarget organisms, and environmental and health hazards) have necessitated the search for and use of effective but safer insect control agents. Unfortunately, few microbial insect control agents have been as successful commercially as chemical insecticides because of difficulties with efficacy and production. Only *Bacillus thuringiensis* has been amenable to in vitro production on a relatively large scale. *B. popilliae* and the few insect viruses approved by the Environmental Protection Agency (EPA) for use as biological insecticides are of limited use because of the lack of in vitro methods of mass production. Fundamental research along alternative avenues of crop protection—such as chromosome, bacteriophage, plasmid, or recombinant DNA manipulation—could lead to the resolution of problems in efficacy and in vitro fermentation encountered with some strains of bacteria used as insect control agents.

THE RELEVANCE OF GENE MANIPULATION STUDIES WITH ENTOMOPATHOGENIC BACTERIA

Envisaged in genetic manipulation of entomopathogenic bacteria are possible applications to expand the spectrum of insect hosts of existing insect pathogens; develop new, more potent strains of pathogens; remove undesirable characteristics from pathogens when needed; improve the physiological tolerance and epidemiological properties of these bacteria; and facilitate the in vitro commercial production of the more fastidious insect pathogens by expanding the range of in vitro substrates upon which they can be grown.

The generally suggested scientific approach to this research is a molecular one comprising a wide range of current biochemical methodology in transformation, plasmid function, and recombinant deoxyribonucleic acid (DNA) research. Strong emphasis should be placed on techniques involving chromosomal and plasmid DNA isolation and characterization; nutritional media, conditions, and biochemical/physiological parameters necessary to develop cell competency and DNA-mediated transformation into selected strains of entomopathogenic bacteria; and construction of recombinant DNA vector systems by using selected gene manipulation enzymes. Within this framework the following questions should be resolved by gene manipulation of these bacteria.

What parameters are necessary for transformation of foreign DNA?

What is the significance of the profiles of extrachromosomal DNA elements in entomopathogenic bacteria to their pathogenic properties?

Is the production of toxic parasporal crystals in the crystalliferous bacteria under the sole control of plasmid DNA(s)?

Will natural and recombinant DNA plasmids be expressed in transformed strains?

What is the manner of expression?

Can transformation of entomopathogenic bacteria with nonchimeric and chimeric plasmids improve their spectrum of activity, pathogenicity, and in vitro fermentation ability?

What are the physiological, biochemical, and pathological properties of new strains (emphasis to be placed on safety, activity spectrum, mode of action, genetic stability, and fermentation capability)?

Is the failure of the milky spore disease organisms to sporulate in artificial media a result of metabolic lesion(s) due to long association as a pathogen growing in hemolymph of Japanese beetle larvae?

Indeed, virtually nothing is known about the mechanism of specificity of different strains of B. thuringiensis for pest insects or about the molecular events involved. The mechanism might function through highly specific plasmid gene variations coding the δ-endotoxin. A clearer insight into the expression of plasmids could profoundly contribute to answering such basic questions as: What is the origin of genetic information coding construction of toxic parasporal crystals? What is the function of the numerous plasmids found in various strains of entomopathogenic bacteria? and What is the molecular mechanism of symptom expression? The abundance and variety of extrachromosomal elements in these bacteria should allow their isolation and purification in quantities sufficient for detailed study of their structure and biological properties and possibly for genetic manipulation. Such manipulation of promising plasmids by simple transformation or recombinant DNA techniques might ultimately improve the efficiency, pathogenicity, and commercial production of entomopathogenic bacteria.

An important spin-off of this fundamental research might be the refinement of techniques that could, for example, be directly applied in research of insect pathology and biological control and that might benefit agriculture worldwide. A better understanding of the origins, functions, biosynthesis, expression, and use of plasmid molecules in recombinant DNA studies with entomopathogenic bacteria could lead to improvements in this new arsenal of insecticides. Furthermore, this research might substantially enhance our chances of understanding the molecular basis of vegetative growth, sporulation, and pathogenicity. The results would also be directly applicable to control of biological plant stress from damage by insects to which most crops are subjected and might lead to more efficient and safer biological control of pest insects. Lastly, results of this research might demonstrate the usefulness of entomopathogenic bacteria as alternate host-vector systems for a variety of recombinant DNA studies.

ATTRIBUTES OF THE ENTOMOPATHOGENIC BACTERIA MODEL SYSTEM

A number of positive attributes of some species of entomopathogenic bacteria used as model systems in recombinant DNA manipulations merit emphasis. First, many of these bacteria, especially B. popilliae and varieties of B. thuringiensis, are "soil" bacilli that are nonpathogenic to nontarget organisms but are primarily species- or genus-specific insect pathogens, and several have been tested for safety to vertebrate and invertebrate nontarget organisms, as will be discussed later. Second, use of bacteriophage vectors to mediate generalized transduction and cotransduction of linked markers for chromosomal mapping has been demonstrated. Third, these bacteria are agriculturally important as biological control agents with potential to replace a number of hazardous chemical insecticides.

In contrast, however, are a number of negative attributes. First, chromosomal/plasmid mapping, the genetics, and the physiology of plasmids and bacteriophages are ill-defined in this group of bacteria. Second, high frequency, specialized transformation, and transduction are not well known as means of gene enrichment. Third, plasmids described to date from these bacteria may be cryptic, a hindrance in selection of transformed strains. Fourth, disabled strains may have to be developed to meet the requirements of the National Institutes of Health (NIH) guidelines for recombinant DNA research. However, if the researcher is not combining potentially hazardous DNA or hazardous heterologous markers, or in cases where DNA from *B. thuringiensis* or *B. popilliae* or from other safe entomopathogenic species are being amplified in the same or other entomopathogenic species proved safe to nontarget organisms, we do not see the necessity of restriction to disabled strains, especially in view of the objectives of the research described earlier. In any event, at least with *B. thuringiensis*, *spo⁻ cry⁺*, *spo⁻ cry⁻*, and *spo⁺ cry⁻* strains have been developed and isolated for use in many experiments. Most of these objections presumably will yield with increased investigations of these model systems.

THE NATURE OF ENTOMOPATHOGENIC BACTERIA

Beneficial bacterial pathogens of insects may be defined as those bacteria that are pathogenic to specific pest insects. Included in this group are *B. thuringiensis* (more than 20 varieties), *B. popilliae*, *B. lentimorbus*, *B. euloomarahae*, *B. fribourgensis*, *B. sphaericus*, *B. moritai*, certain strains of *B. cereus*, *Clostridium brevifaciens*, and *C. malacosomae*. The eight *Bacillus* organisms are either now used commercially or show the greatest potential for use. Other entomopathogenic bacterial species have been found in the genera *Pseudomonas*, *Aerobacter*, *Cloaca*, and *Proteus*, but these have not been seriously considered for use in biological control of pest insects because of their nonspecificity. Table 15.1 lists the entomopathogenic bacteria

TABLE 15.1
Entomopathogenic Bacteria with Greatest Potential for
Use in Widespread Biological Control of Insect Pests

Organism	Host(s)	Mechanism of Action
Bacillus popilliae, *lentimorbus*, *fribougensis*, *euloomarahae*[a]	Coleopteran larvae, especially Japanese beetle and European chafer	Causes milky spore disease, septicemia

B. thuringiensis varietal types such as *kurstaki, sotto, dendrolimus, alesti, entomocidus*[b]	Numerous lepidopteran pest larvae	Parasporal crystal toxin, disruption of midgut wall, paralysis
B. cereus (thuringiensis) var. *juroi*[c]	Mosquito larvae (*Culex, Aedes*)	Cuboidal crystal toxin, disruption of gut wall
B. thuringiensis BA 068[d]	Mosquito larvae (*Aedes aegypti, Culex*), some Lepidoptera	Parasporal crystals (bicrystalliferous) and bacteremia
B. thuringiensis var. *israelensis*[e]	Mosquito larvae (*Aedes aegypti, Anopheles, Culex*)	Parasporal crystal toxin, disruption of midgut wall
Special HD strains of *B. thuringiensis (kurstaki, thuringiensis, tolworthi, galleriae, morrisoni)*[f]	Mosquito larvae (*Aedes* and *Culex*), some Lepidoptera	Parasporal crystal toxin, septicemia
B. moritai[c]	Housefly, stable fly, seed corn maggot	Toxic principle of in vivo growing bacteria, inhibits larval development
B. sphaericus (SSII-1)[g]	Mosquito larvae (*Culex, Culiseta, Aedes*), Clear Lake gnat	Toxin-mediated, located in outermost cell wall layer, disrupts midgut cells
B. alvei-circulans[h]	Mosquito larvae (*Culex*)	Heat-labile soluble toxin, action unknown
B. cereus[b]	Hymenoptera, Lepidoptera, Coleoptera	Septicemia
Clostridium malacosomae[i]	Tent caterpillar	"Brachyosis," bacteremia of gut, toxic paralysis
C. brevifaciens[i]	Tent caterpillar	"Brachyosis," bacteremia of gut, toxic paralysis
Clostridium sp (?)[i]	Essex skipper larvae	"Brachyosis," bacteremia of gut, toxic paralysis

[a]Beard 1956; Dutky 1940; Willie, 1956.
[b]Heimpel 1967a, 1967b; Heimpel and Angus 1963.
[c]Fujiyoshi 1973.
[d]Reeves and Garcia 1971.
[e]de Barjac 1978a,1978b.
[f]Hall et al. 1977.
[g]Davidson, Singer, and Briggs 1975; Kellen et al. 1965; Sikorowski and Madison 1968; Singer 1974.
[h]Singer 1973.
[i]Bucher 1961.
Source: Compiled by the authors.

that have the greatest potential for use in widespread biological control of insects pests and that may be amenable to genetic manipulation.

Mode of Action and Host Range

All entomopathogenic spore-forming bacteria produce endospores that allow them to persist in a dormant or quiescent state outside the host. Upon ingestion by a susceptible host the spores may germinate in the gut. In causative organisms of milky disease (*B. popilliae, B. lentimorbus, B. fribourgensis,* and *B. euloomarahae*), which primarily affect larvae of beetles in the order Coleoptera, the vegetative cells produced by the germinating spores enter the hemocoel, multiply rapidly, destroy certain tissues, and soon fill much of the cavity. Before the host dies, thick-walled refractile spores are formed, which appear white through the integument; thus the name milky disease. After the host dies, it disintegrates, and the spores are released into the soil. In the genus *Clostridia* the mode of action differs in that these obligate pathogens multiply only in the gut and do not invade the hemocoel.

The crystalliferous spore formers (varieties of *B. thuringiensis*), in addition to forming endospores, produce a proteinaceous parasporal crystal in the sporangium at the time of sporulation. The crystal contains an endotoxin capable of paralyzing the gut of most pest lepidopteran larvae (Heimpel 1967a; Heimpel and Angus 1963) and some pest mosquito larvae, depending on the *B. thuringiensis* strain (de Barjac 1978a,1978b). This material has been labeled δ-endotoxin (Heimpel 1967a; Heimpel and Angus 1963); susceptible insects are killed by the toxic crystals (Heimpel and Angus 1963).

Faust, Travers, and Heimpel (1979) have attempted to correlate the biochemical and histological temporal sequence in *B. thuringiensis* δ-endotoxin intoxication of insects. As shown in Table 15.2, glucose uptake is stimulated in the larval midgut of *Bombyx mori* (L.) within 1 minute after susceptible insects ingest the toxic crystals. Stimulation is maximum within 5 minutes, and glucose uptake ceases in 10 minutes. Control of cell permeability breaks down between 10 and 15 minutes as levels of K^+ and other ions increase in the hemolymph. However, K^+, Na^+, Ca^+, and Mg^{2+} do not accumulate in gut tissue during the first 10 minutes after toxin is given *per os*, and no change is detected in the rate at which K^+ turns over in the gut tissue. After 12 minutes' exposure in *Pieris brassicae* (L.), the fine structure of the epithelial cells begins to disrupt, and microvilli containing mitochondria swell rapidly. The contents of the mitochondria seem to dissolve during this continuous enlargement, leaving spherical cavities. The cristae of mitochondria disintegrate and dissolve, leaving an empty cell. After ap-

TABLE 15.2
Correlation of Biochemical and Histological Temporal Sequence of *Bacillus thuringiensis* δ-Endotoxin Intoxication in Certain Lepidopteran Larvae

Time (minutes)	Biochemical Effects	Histological Effects
0–1	Degradation of *B. thuringiensis* δ-endotoxin[a,b] Glucose uptake stimulated; amino acid, carbonate ion, mono- and divalent cation transport unchanged[c,d] Acceleration O_2 uptake in H^+ transport system[e,f]	None
1–5	Maximum glucose uptake[c,d] Maximum acceleration of O_2 uptake; uncoupling of oxidative phosphorylation, and cessation of ATP production[e,f]	None
5–10	Cessation of glucose uptake; no changes in K^+, Na^+, Ca^+, Mg^+ in gut tissue[c,d]	Irregular swelling and distortion of gut microvilli[g]
10–20	General breakdown in permeability control; K^+ and other ions increase in hemolymph; loss of osmotic integrity[d,h,i] Rise in silkworm blood pH; initial paralysis of gut[j]	Epithelial gut cells swell, protrusions develop, endoplasmic reticulum disrupts; mitochondria increase in size, cristae disintegrate[g,h,l]
After 20	Paralysis of silkworm gut[m,n,o]	Epithelial cells disrupt, lyse, and slough off into lumen[j,k,l,p,q]

[a]Fast and Videnova 1974.
[b]Fast 1975.
[c]Fast and Donaghue 1971.
[d]Fast and Morrison 1972.
[e]Faust, Travers, and Hallem 1974.
[f]Travers, Faust, and Reichelderfer 1976.
[g]Ebersold, Luthy, and Mueller 1977.
[h]Fast and Angus 1965.
[i]Louloudes and Heimpel 1969.
[j]Heimpel and Angus 1959.
[k]Sutter and Raun 1967.
[l]Angus 1970.
[m]Hannay 1953.
[n]Angus 1954.
[o]Young and Fitz-James 1959.
[p]Angus and Heimpel 1959.
[q]Hoopingarner and Materu 1964.
Source: Compiled by the authors.

proximately 25 minutes the columnar cells balloon or extrude noticeably and severely disrupt the gut wall as they burst.

Valinomycin was reported to cause similar effects (Angus 1968). The effect of valinomycin on mitochondria is to uncouple respiration from phosphorylation and stimulate adenosinetriphosphatase (ATPase) activity. Valinomycin also affects certain plasma membranes, causing altered levels of K^+ (Andreoli, Tieffenberg, and Toteson 1967; Toteson et al. 1967); however, Fast and Morrison (1972) reported that change in K^+ concentration was not significant in gut tissue within the first 10 minutes after ingestion of the δ-endotoxin. If the primary effect of the toxin were on the cell surface membrane, then cation flux would precede an increase in glucose uptake (cause and effect). The temporal sequence of events does not favor a hypothesis based on the primary interaction of the δ-endotoxin with the membrane of the epithelial cells. Recent evidence (Faust, Travers, and Hallem 1974; Travers, Faust, and Reichelderfer 1976) suggests that the δ-endotoxin stimulates mitochondrial oxygen uptake and inhibits oxidative phosphorylation. Loss of adenosine triphosphate (ATP) production, caused by the action of the δ-endotoxin as an uncoupler, would lead to metabolic imbalance.

Faust, Travers, and Hallem (1974) and Travers, Faust, and Reicheiderfer (1976) offered a theory that integrated the temporal sequences of the histological and biochemical effects of the δ-endotoxin: stimulation of respiration (O_2 uptake) by the δ-endotoxin could stimulate glucose uptake during the first 5 minutes of intoxication in susceptible insects. That is, high demand for reducing potential in a nonconservative electron transport system would increase the demand for glucose in the insect catabolic process, causing the sudden influx of glucose into the midgut cells. The loss of ATP production caused by uncoupling would result in the cessation of glucose metabolism by the midgut cells. As the energy levels fell, a loss of osmotic integrity and regulation would cause the cells to take up water, balloon, and lyse.

In addition to the crystal at least three other substances toxic to insects may be produced by various strains of crystalliferous bacteria. Following the nomenclature proposed by Heimpel (1967a) these are: α-exotoxin, an enzyme of the growing bacterium; phospholipase C, which breaks down essential phospholipids in insect tissue; β-exotoxin, a small, water-soluble, heat-stable nucleotide that is a structural analogue of ATP, which inhibits DNA-dependent ribonucleic acid (RNA) polymerase (Sebesta and Horska 1968) and kills larvae and pupae of Diptera and some Lepidoptera; and γ-exotoxin, an unidentified phospholipase that affects phospholipids, probably releasing fatty acids from the molecule. The β-exotoxin has been demonstrated as toxic to mammals; however, a number of varieties of B. *thuringiensis* do not produce the β-exotoxin (Faust 1975), and these varieties are now used in commercial formulations. *B. thuringiensis* and its varieties have been tested successfully against more than 137 insect species form the orders Lepidoptera, Hymenoptera, Diptera, and Coleoptera (Heimpel 1967a). Most of these pathogenicity tests have been conducted in the laboratory. The most susceptible insects are those lepidopteran larvae with moderately alkaline gut contents (pH 9.0–10.5) and enzymes that dissolve the crystals and release the toxin (Angus 1956; Angus and Heimpel 1959; Heimpel and Angus 1959).

The non-spore-forming entomopathogenic bacteria are a heterogeneous group that usually infect insects only under extraordinary circumstances. Although they inhabit the gut of the host, they do not multiply readily there and probably do not produce enzymes or toxins in sufficient quantities to cause damage or aid invastion of the hemocoel. They do, however, multiply in the hemocoel of the insect once they have invaded it (Bucher 1963). The non-spore-forming bacteria attack a wide range of hosts from the Acarina and from the insect orders Coleoptera, Diptera, Hymenoptera, Isoptera, Lepidoptera, and Orthoptera (Bucher 1963).

B. cereus, a spore-forming facultative pathogen, also produces enough phospholipase C to damage gut cells of susceptible species, thus aiding entry of the bacterium into the hemocoel (Heimpel 1955; Heimpel and Angus 1963). It is a common, widely distributed soil saprophyte with an extensive host range including insect species from the orders Coleoptera, Hymenoptera, and Lepidoptera (Heimpel and Angus 1963). Its host range is limited in part by the pH of the gut (a highly alkaline gut content can inactivate the toxic exoenzyme phospholipase C) (Heimpel 1955).

Clostridial pathogens were originally isolated from *Malacosoma californicum pluviale* (Dyar) (Bucher 1957). Other than experimental infections in other tent caterpillars and the possible presence of these bacteria in *Thymelicus lineola* (Ochsenheimer) (Heimpel and Angus 1963), little is known about their host range.

Methods of Propagation, Factors that Influence Effectiveness, and Relevance of Recombinant DNA Technology

The obligate pathogens among species of *Bacillus* and *Clostridium* are fastidious organisms that can be propagated in artificial media only with extreme difficulty. Most other bacterial pathogens are much less fastidious and can be readily propagated on simple bacteriological media, which is a reason why the facultative bacterial pathogens are likely candidates for microbial control.

Unfortunately, several obstacles block commerical production and widespread use of many biological control agents as replacements for the well-known and hazardous chemical insecticides. For example, wide-scale biological control of the Japanese beetle (*Popillia japonica* Newman), the European chafer (*Amphimallon majalis* Razoumowsky), and other susceptible scarabaeid grubs, all of which are major pests of lawns, pastures, and other plant life in many parts of the world, could be considerably facilitated by the development of an in vitro industrial method to produce spores of *B. popilliae*, the causative agent of milky disease in these insects. Spore preparations of *B. popilliae* are produced commercially by collecting living larvae form infested soil, injecting each with the disease organism, incubating them until the hemolymph becomes filled with spores, then grinding and mixing them with an extending material such as talc. However, this procedure produces a relative expensive, low-yielding product incapable of meeting the requirements for adequate mass control of these pest insects.

Despite approximately 35 years of research and voluminous data concerning the pathogenicity, physiology, and biochemistry of *B. popilliae*, in vitro methods for large-scale production of infective spores, the form necessary for long-term survival of the infectious agent and control of the insect, have not been developed. The low sporulation rate of selected strains necessitates long incubation periods with rare production of sporulating vegetative cells, and such spores are only minimally infective to Japanese beetle larvae *per os*. This negative feature is also present in clostridial entomopathogens.

Fundamental research along alternative avenues of crop protection, such as transformation and/or recombinant DNA formed from genes of *B. thuringiensis*, an inexpensively grown and readily sporulating insect pathogen, could lead to resolution of in vitro fermentation and sporulation problems. Subsequent fundamental research on the pathogenicity, safety, and physiological and biochemical properties of newly developed *B. popilliae* strains may allow mass in vitro industrial production of highly selective and virulent strains.

Another organism, *Bacillus sphaericus*, is a potentially powerful biological control agent for mosquitoes. The pathogenicity of this organism for mosquito larvae seems to reside in its capacity for producing an endotoxin specific for the midgut of the host (Davidson, Singer, and Briggs 1975). Recombinant DNA incorporation of the genes responsible for production of this toxin and transformation of them into *B. thuringiensis*, the biological lepidopteran insecticide now produced at a price competitive with chemical insecticides, could yield a biological control agent of broader spectrum affecting several orders of economically important insect pests. Although broader based, this new strain would be selective, an attribute shared by few chemical insecticides. The development and use of such strains could decrease the use of nonselective chemical insecticides. In fact, a recently isolated strain of *B. thuringiensis* (var. *israelensis*) produces parasporal crystals (δ-endotoxin) toxic for such pest mosquito larvae as *Aedes aegypti* (L.) and *Anopheles stephensi* Lister but is not toxic to larvae of Lepidoptera (de Barjac 1978b). Combination of these pathogenic entities into one strain would give industry a very valuable product. *B. moritai*, pathogenic for houseflies, stable flies, and the seed corn maggot (Fujiyoshi 1973), produces a toxin that inhibits larval growth; similar research approaches to those described here may also prove advantageous.

General Safety Considerations

One of the first entomopathogenic bacteria used extensively in the field was a causal agent of milky disease, *B. popilliae*, isolated and described by Dutky (1940). This organism has been used on the Eastern Seaboard since the early 1940s to control the Japanese beetle *Popillia japonica*. The only producers of this bacterial product are the Fairfax Biological Laboratory* at Clinton Corners, N.Y., and Reuter Laboratories, Inc., at Haymarket, Virginia. The materials produced by Fairfax Biological Laboratory are manufactured under the trade names of Japanese Beetle Doom (USDA Reg. no. 403-9) and Japidemic (USDA Reg. no. 403-14). Reuter Laboratories manufacture their product under the trade name of Milky Spore (EPA Reg. no. 36488-1). Numerous tests showed that the milky disease bacteria are innocuous to turf, invertebrates, and vertebrates, including man (Anonymous 1977; Bailey 1976; Bulla, Rhodes, and St. Julian 1975; Heimpel 1971; Heimpel and Hrubant 1973; Ignoffo 1973). The protocols for these tests were informally discussed, accepted, and recommended by officials of the EPA and the U.S. Department of Agriculture (USDA) and were conducted by qualified investigators in accordance with these protocols.

The following tests were conducted to ascertain the safety of *B. popilliae* to man and other mammals: safety evaluation by repeated oral administration of *B. popilliae* to rats and monkeys; clinical examination of production personnel exposed to *B. popilliae*; medical and serological examinations of humans engaged in production of *B. popilliae*; microbiological examination of commercial preparations of *B. popilliae*; studies of the specificity of *B. popilliae* to beetle larvae and its safety to vertebrates, including humans; tests for acute dermal toxicity in guinea pigs with a spore preparation of *B. popilliae*; a study of potential acute eye irritation in rabbits with a spore preparation of *B. popilliae*; and tests to determine the ability of *B. popilliae* to grow at mammalian or avian body temperatures.

---

*Mention of a company or proprietary product does not constitute an endorsement by the U.S. Department of Agriculture, the University of Maryland, or the University of Massachusetts.

The data supported the relief sought by the petition — namely, exemption from the requirement of a tolerance. *B. popilliae* is so specific that even its misuse cannot threaten the safety of humans or other animals; it has been proved completely innocuous wherever and at whatever levels used. This milky disease organism has never been shown to infect anything but Japanese beetles and certain closely related beetles. Application of up to 2,000 pounds per acre of milky disease spore powder to pastures caused no discernible harm to turf. Temperature is the defense of vertebrates against infection by milky disease bacteria; the maximum temperature at which *B. popilliae* multiplies is below the minimum body temperature of domestic vertebrates. Tests also have shown that this organism does not germinate in the digestive tract or persist in the feces of vertebrates.

B. thuringiensis has been commercially available in the United States since 1958 and is widely used to control caterpillar pests of food crops, fiber crops, and forests. Both laboratory and commercial preparations of *B. thuringiensis* have been evaluated since 1957 for specificity and possible toxicity-pathogenicity to vertebrates. International Minerals and Chemical Corporation (IMC) produced the first commercial preparations of *B. thuringiensis* (Thuricide®) in July 1957. The toxicology of Thuricide was included in a petition to the Food and Drug Administration (FDA) that resulted in a temporary exemption from tolerance in December 1958 and a full exemption from tolerance for use on food and forage crops in April 1960. The Canadian Department of Agriculture granted a similar registration for Thuricide in November 1961. Another petition, which contained safety evaluation data, was submitted in June 1959 by Nutrilite Products, Inc. (NPI) for their preparation of *B. thuringiensis* (Biotrol-BTB®). The petition on Biotrol-BTB presented results on the following studies: intraperitoneal injection in the mouse, guinea pig, rabbit, swine, and chick; serial blood passage after intraperitoneal injection into mice; and ingestion by the rat, chick, and human. Subsequent studies by NPI included subacute feeding tests in birds and mammals. Another commercial preparation of *B. thuringiensis* (var. *kurstaki*), Dipel®, produced by Abbott Laboratories, was developed and registered in 1970 (EPA Reg. no. 275-18-AA). Preparations were evaluated for their safety to fish, birds, and mammals, including man. Initial studies were designed to establish that *B. thuringiensis* and its varieties were neither pathogenic, allergenic, nor persistent in mammals and to answer questions concerning the possibility that *B. thuringiensis* might mutate or be selected for pathogenicity to humans. Acute and subacute dosage studies were conducted to establish requirements for registration of commercial preparations of *B. thuringiensis*. The results demonstrated that *B. thuringiensis* has no known adverse effects on man, pets, birds, fish, earthworms, beneficial insects, or plants (Anonymous 1978).

Other studies revealed no evident differences of snails, *Forficula*, myriapodes, or wood lice after treatment (Benz and Altwegg 1975). Millions of pounds of Dipel have been used without a single report of human toxicity or environmental damage. Repeated tests have shown that *B. thuringiensis* does not inhibit plant growth and is nonphytotoxic to more than 140 species of plants. This bacterium has been proven harmless to vertebrates and invertebrates in terrestial, marine, and freshwater environments even at test dose levels of 1,000 to 2,000 times the normal rate. Laboratory studies indicated that ingested *B. thuringiensis* is eliminated from test animals within a few hours after ingestion (Anonymous 1978). Other studies have demonstrated that vegetative cell counts of *B. thuringiensis* decreased by 90 percent within 4 hours when bacterial suspensions were introduced into the rumens of cattle, with commercial spore preparations decreasing after 24 hours (Adams and Hartman 1965). No germination of spores in the rumens was noted in that study. Studies of the survival of *B. thuringiensis* in the digestive tracts and feces of mammals and birds showed similar results, although spores survived passage (Smirnoff and MacLeod 1961).

No evidence of acute or chronic toxicity in rats, guinea pigs, mice, swine, humans, or other mammals has been found. *B. thuringiensis*, either in the form of the commercial product (Dipel) or as naturally occurring *B. thuringiensis*, has no harmful effects on the environment for several reasons, all of which are characteristic of the bacteria (Anonymous 1978; Forsberg et al. 1976) First, *B. thuringiensis* numbers in the soil are limited by metabolic requirements; it is apparently unable to compete effectively with other soil bacteria and, unlike *B. cereus*, is seldom isolated and reported from soil but is isolated and reported from diseased larvae. Cadavers of insects killed by *B. thuringiensis* contain vegetative cells but no spores or crystals; the principal agents of mortality rarely build up in the insect population. Second, conditions on the surfaces of living plants are unsuitable for propagation of *B. thuringiensis*. Third, numbers of *B. thuringiensis* in the environment are reduced quickly by ultraviolet light, even after application of concentrated commercial formulations; the number of recoverable *B. thuringiensis* spores usually falls to the background level in 2 or 3 days. Fourth, *B. thuringiensis* is toxic to target insects only when ingested. And fifth, what little prolific multiplication there is of *B. thuringiensis* in insect larvae occurs just before or after the death of the larvae (Prasertphon, Areekul, and Tanada 1973).

Extensive experimental work on the safety of *B. thuringiensis* (and certain strains of *B. cereus*) as insect control agents has also been reported by Fisher and Rosner (1959), Ignoffo (1973), Heimpel (1971), Krieg and Franz (1959), Lamanna and Jones (1963), Lemoigne (1956), and Steinhaus (1951). Examined in these studies were *B. thuringiensis* virulence in mice, persistence in blood of mammals, pathogenicity by parenteral administration, inhalation toxicity in mice, allergenicity in guinea pigs, and inhalation and ingestion by human volunteers.

Bacillus moritai, a pathogen for houseflies and mosquitoes, is manufactured by private industry in Japan. Fujiyoshi (1973) investigated the effectiveness of this product in the field and its safety for warm-blooded animals. *B. moritai* has been proved nonpathogenic for silkworms, honey bees, mice, rats, rabbits, birds, fish, pigs, and cattle, and studies of malignancy and carcinogenicity proved that it is harmless to humans.

When considering selected entomopathogenic bacteria as possible hosts for recombinant DNA, one should keep in mind that neither *B. thuringiensis* nor *B. popilliae* establish themselves in the healthy bowel or multiply in the alimentary tract of humans, although spores remain viable during passage through the intestinal tract. Although *B. popilliae* and *B. thuringiensis* survive only in the spore stage in the environment, they affect specific pest insects, and *B. thuringiensis* normally does not multiply in the target insect. Neither bacterium is capable of spreading from animal to animal or plant to plant except in target insects. They do not multiply on body surfaces, intestines, or lungs, nor do they penetrate animal cells or spread through animal bodies except those of target insects. They produce toxins that affect only specific pest insects (except the few varieties of *B. thuringiensis* that produce β-endotoxin, which are not approved for use by the EPA). They do not resist normal body defense mechanisms or establish themselves as permanent residents in humans or other nontarget organisms.

FEATURES RELATING TO GENE MANIPULATION

The entomopathogenic bacteria are primarily species or genus specific, and several strains have been tested for their safety to vertebrate and invertebrate nontarget organisms. It is suggested by the authors that increased emphasis in recombinant DNA research be focused on those bacteria with safety advantages over bacteria currently used as recombinant DNA tools. The use (and approval where appropriate) of plasmid or bacteriophage vectors is a primary consideration in these research efforts.

The Incidence of Plasmids in Naturally Occurring Entomopathogenic Bacteria

Plasmids have been identified in many bacteria, and a wide variety of specific biochemical functions (fertility, resistance to antibiotics, production of bacteriocins, production of toxins, et cetera) have been attributed to these genetic elements (Cohen 1976; Helinski 1973; Helinski and Clewell 1971). Extrachromosomal DNA elements also have been found in spore-forming bacteria such as *B. pumilus* (Lovett 1973; Lovett and Bramucci 1975; Lovett and Burdick 1973; Lovett, Duvall, and Keggins 1976), *B. subtilis* (Lovett and Bramucci 1974, 1975; Tanaka and Koshikawa 1977; Tanaka, Kuroda, and Sakaguchi 1977), *B. megaterium* (Carlton and Helinski 1969), *B. thuringiensis* (Debabov et al. 1977; Ermakova et al. 1978; Faust et al. 1979; Galushka and Azizbekyan 1977; Miteva 1978; Stahly, Dingman, Bulla, and Aronson 1978; Stahly, Dingman, Irgens, Field, Feiss, and Smith 1978), and *B. popilliae* (Dingman and Stahly 1979; Faust et al., 1979). Table 15.3 lists the number and estimated size of indigenous extrachromosomal DNA elements in six varieties of *B. thuringiensis* and in *B. popilliae.*

There is some evidence that plasmid(s) may be involved in the synthesis of the parasporal crystals (*B. thuringiensis* δ-endotoxin) responsible for the pathogenicity of *B. thuringiensis* to insects (Debabov et al 1977; Ermakova et al. 1978; Galushka and Azizbekyan 1977; Stahly, Dingman, Bulla, and Aronson 1978). For example, Stahly, Dingman, Bulla, and Aronson (1978) examined crystalliferous and acrystalliferous strains of *B. thuringiensis* var. *kurstaki* and found that crystalliferous strains contained at least six extrachromosomal DNA molecules, ranging from 1.32×10^6 to 47.13×10^6 daltons. All nontoxic acrystalliferous mutants lacked the complete array of at least six plasmids present in the wild type, an implication of a relationship between presence of plasmid(s) and toxicity. Furthermore, the very high frequency with which acrystalliferous mutants appeared suggested involvement of an unstable genetic element such as a plasmid.

Miteva (1978) examined 14 strains of *B. thuringiensis* and 6 strains of closely related *B. cereus* for the presence of plasmid DNA. Results of electrophoretic analysis demonstrated that all of these *B. thuringiensis* strains possessed covalently closed circular double-stranded DNA. In *B. cereus* the presence of plasmid DNA was established in three out of six strains. However, the plasmids isolated from some asporogenous and acrystalline mutants of *B. thuringiensis* appeared no different from the parent strains. The fact that the asporogenous and acrystalline mutants retained both their plasmids and their ability to form crystals does not exclude the possibility that plasmids of *B. thuringiensis* take part in the genetic determination of crystal formation. The loss of protein crystals in this study might have been the result of a mutation.

Ermakova et al. (1978) demonstrated with *B. thuringiensis* var. *galleriae*, grown in selective media that inhibited crystal formation, that no extrachromosomal DNA could be isolated from the cells. The results suggested a correlation between the presence of plasmid DNA, the formation of crystals, and the level of intracellular protease activity. The results of medium-

TABLE 15.3

Number and Estimated Size of Extrachromosomal DNA Elements in Varieties of *Bacillus thuringiensis* and *Bacillus popilliae* ($\times 10^6$ daltons)

kurstaki[a]	sotto[a]	finitimus[a]	galleriae[b]	galleriae[c]	alesti anduze[d]	kurstaki[e]	B. popilliae[a]	B. popilliae[f]
>50		>50(2)			44.58	47.13		
~45	~23.5				36.72	30.10		
~29.9					18.12	9.64		
~17.1					9.96			
7.4			12.0	10.9	8.67	5.45		10.5
4.2			10.0	10.0	8.10	4.90		10.0
3.9			6.0	5.9	7.29	1.32	4.45	5.37
3.6					6.29			
1.1		0.98			5.32			
0.87					4.75			
0.80	0.80	0.79			3.80		0.58	
0.74	0.62				2.60			

[a]Faust, Spizizen, Gage, and Travers 1979.
[b]Debabov et al. 1977.
[c]Ermakova et al. 1978.
[d]Stahly, Dingman, Ingens, Field, Feiss, and Smith 1978.
[e]Stahly, Dingman, Bulla, and Aronson 1978.
[f]Dingman and Stahly 1979.

Source: Compiled by the authors.

shift experiments of their research support their hypothesis that plasmid DNA may have a chromosomal origin, perhaps the result of specific excision and amplification of certain chromosomal DNA segments. Previously reported experiments with *B. megaterium* plasmids (Carlton 1976) are consistent with this explanation. Galushka and Azizbekyan (1977) also demonstrated that no extrachromosomal DNA elements could be detected after several *B. thuringiensis* strains were cured with ethidium bromide or cultured under extreme conditions (increased temperature, alkaline pH of medium) and selected for acrystalline strains. Debabov et al. (1977) reported similar results. Results from transformation experiments involving acrystalliferous bacilli should definitively prove whether extrachromosomal DNA elements control *B. thuringiensis* δ-endotoxin production.

Other researchers have also reported plasmid DNAs in *B. thuringiensis* and *B. popilliae*. Stahly, Dingman, Irgens, Field, Feiss, and Smith (1978) examined extrachromosomal DNA by electron microscopy in *B. thuringiensis* var. *alesti* and identified 12 size classes of molecules ranging from 2.60×10^6 to 44.58×10^6 daltons. Debabov et al. (1977) examined *B. thuringiensis* var. *galleriae* and found three kinds of circular DNA molecules with sizes of 6.0×10^6, 10.0×10^6, and 12.0×10^6 daltons. Similarly, Ermakova et al. (1978) examined *B. thuringiensis* var. *galleriae* and also found three plasmids with molecular weights of 5.9×10^6, 10.0×10^6, and 10.9×10^6 daltons. Galushka and Azizbekyan (1977) examined *B. thuringiensis* var. *galleriae*, var. *thompsoni*, var. *tolworthi*, and var. *finitimus* and also found plasmid DNA in these varieties but reported no molecular weight sizes in their results. It appears, therefore, that the varietal types of *B. thuringiensis* exhibit different plasmid profiles.

Faust et al. (1979) recently examined four strains of entomopathogenic bacteria for extrachromosomal DNA molecules. The basic characteristics of the larger isolated elements are not unlike those of representative plasmids isolated from members of other genera of bacteria (Clowes 1972). *Bacillus thuringiensis* var. *kurstaki* contained 12 elements that banded on agarose gels; these elements ranged from 0.74×10^6 to $>50 \times 10^6$ daltons, and 3 were very large extrachromosomal DNA elements. *B. thuringiensis* var. *sotto* contains one very large extrachromosomal DNA element with a molecular size of about 23.5×10^6 daltons and two smaller elements of 0.80×10^6 and 0.62×10^6 daltons. *B. thuringiensis* var. *finitimus* harbors two very large DNA elements corresponding to $>50 \times 10^6$ daltons and two elements of relatively small size (0.98×10^6 and 0.79×10^6 daltons). *B. popilliae* contains no large extrachromosomal DNA elements but does contain 2 small elements corresponding to 4.45×10^6 and 0.58×10^6 daltons. Further genetic and biochemical studies are obviously necessary to determine the biological functions of extrachromosomal DNA elements in *B. thuringiensis* and *B. popilliae* and to definitely determine which elements are indeed autonomous replicons that may be useful in genetic manipulation to improve their effectiveness.

Unfortunately, most of the described plasmids in *Bacillus* are cryptic elements (Lovett and Bramucci 1975; Lovett, Duvall, and Keggins 1976; Tanaka, Kuroda, and Sakaguchi 1977) lacking genetic markers and are unsuitable for selection of transformed colonies. This may also be the case with plasmid DNA in entomopathogenic bacteria. However, with recombinant DNA techniques, one could conceivably splice markers such as antibiotic resistance genes into cryptic plasmids of *B. thuringiensis* to render them more useful in selection studies.

Bernhard, Schremph, and Goebel (1978) have undertaken a search for plasmids in *Bacillus* species, mainly *B. cereus* and *B. subtilis*, with the intent of characterizing their properties and developing their potential use as vectors for gene cloning. Most of the *B. cereus* strains examined to date contain two or more plasmids with molecular weights ranging from 1.6×10^6 to 105×10^6 daltons. Bacteriocin production and tetracycline resistance could be attributed to a 45×10^6 and a 2.8×10^6 dalton plasmid, respectively, both from *B. cereus*. The plasmid conferring resistance to tetracycline, which was originally isolated from *B. cereus*, could be subsequently transformed in *B. subtilis*, where it was stably maintained. Varietal types of *B. thuringiensis* have been shown to be resistant to streptomycin (Afrikian 1960), penicillin, polymyxin B, nystatin, bacitracin, viomycin (Ignoffo 1963; Krieg 1969), oxytetracycline (Fargette and Grelet 1976), and tetracycline (Fargette, Grelet, and Rapoport 1978; Fargette, Rapoport, and Grelet 1978). Bacteriocin production has also been identified in *B. thuringiensis* (Krieg 1970). The location of genes expressing these characteristics is unknown.

An alternative approach, which we are pursuing, involves the introduction of plasmids from other species. In fact, the first successful report of transfer of foreign plasmids into bacilli involved the RP1 plasmid of *E. coli* transferred to *B. subtilis* (Domardskii et al. 1976; Ehrlich 1977). More recently, Ehrlich (1977) demonstrated that one plasmid encoding tetracycline resistance (pT127) and four plasmids encoding chloramphenicol resistance (pC194, pC221, pC223, and pUB112) could be introduced into *B. subtilis* by DNA-mediated transformation. Ehrlich (1977) also reported that several *Staphylococcus aureus* plasmids can replicate and express antibiotic resistance in *B. subtilis*, a finding of great interest, since transformation and recombinant DNA experiments in the *Bacilli* have previously been hindered by the absence of vectors with selectable markers. Aside from their use for molecular cloning, these plasmids are also potentially useful for studies on plasmid biology, DNA uptake by competent cells, and genetic recombination in the *Bacilli*, especially the entomopathogenic bacteria.

The plasmids from *S. aureus* have been characterized extensively by Gryczan and Dubnau (1978) and by Lovett and coworkers (see Young and Wilson 1978). These plasmids are both maintained and readily amplified by *B. subtilis*. Furthermore, it has been possible to clone the gene(s) in the tryptophan biosynthetic pathway of *B. licheniformis*, *B. pumilus*, and *B. subtilis* in the *S. aureus* pUB110 and to complement the genetic defect in the *trpC2* locus of 168 (Keggins, Lovett, and Duvall 1978). Since the characterization of the plasmids from *S. aureus* has been extensive, they should provide important cloning vectors. The exact number of plasmid copies and the proportion in the extrachromosomal state versus those integrated into the chromosome of *B. subtilis* are not known. Further studies are essential to explore these observations in detail and to develop these cloning vehicles.

The pUB110 plasmid (kan^r, neo^r) from *S. aureus* has been suggested for use in genetic manipulation studies of the entomopathogenic bacteria (Faust 1979). Restriction endonuclease cleavage sites on the *S. aureus* pUB110 plasmid (Gryczan, Contente, and Dubnau 1978) are as follows: $AluI = 5$, $BamHI = 1$, $BglII = 1$, $EcoRI = 1$, $HaeIII = 4$, $HindII = 2$, $HpaII = 4$, and $XbaI = 1$. Restriction endonuclease cleavage sites have also been mapped for the pUB110 plasmid (Gryczan Contente, and Dubnau 1978). In several cases, other *S. aureus* plasmids (pUB101-Fus^r, pK545-Km^r, and pSH2-Km^r) may integrate in part or in toto into the bacterial chromosome and may be useful for integrating additional genes involved in the sporulation process into the *B. popilliae* genome. Use of the *XbaI*, *EcoRI*, *BamHI*, and *BglII* restriction endonucleases to analyze and construct the pUB110 chimeric plasmids is important, since they have been successfully used to insert foreign DNA without loss of antibiotic resistance characters and do not interfere with the genes essential for replication in pUB110 (Gryczan and Dubnau 1978). Before our attempts at transformation with the pUB110 plasmid, we examined 13 serotypes of *B. thuringiensis* and the commercial strain of *B. popilliae* for inherent antibiotic susceptibility/resistance and mutation potential to neomycin and kanamycin. Three varieties of *B. thuringiensis* were found to be doubly resistant, seven varieties were singly resistant (neo^r), and three other varieties were susceptible to both antibiotics (neo^s/kan^s); *B. popilliae* was susceptible to both antibiotics. Estimates of mutation rates revealed that three serotypes developed no resistant mutants to either antibiotic in populations as high as 3.0×10^{10}; seven other serotypes developed no resistance to kanamycin in populations as high as 4.6×10^9 cells. Three other serotypes exhibited mutation rates as high as 1.6×10^{-2} to both antibiotics. We were unable to determine the mutation rate for *B. popilliae*. We now have presumptive evidence that one selected strain of *B. thuringiensis* is transformable with the pUB110 plasmid. We are presently attempting to confirm these findings.

The Incidence of Bacteriophages in Naturally Occurring Entomopathogenic Bacteria

A number of bacteriophages have been reported in entomopathogenic bacteria, and some have been suggested for use in mediating generalized transduction in *B. thuringiensis* (Ackermann, Smirnoff, and Bilsky 1974; Chapman and Norris 1966; Colasito and Rogoff 1969; de Barjac, Sisman, and Cosmao-Dumanoir 1974; Norris 1961; Thorne 1978; Van Tassell and Yousten 1976; Yoder and Nelson 1960). In fact, some examples of cotransduction and linked markers in *B. thuringiensis* have been presented, a demonstration of the feasibility of chromosomal mapping in this organism (Thorne 1978). Two linkage groups were demonstrated. One group included linkage of *trp-1* to *cys-1* and *cys-2* but not to *met-1*, and the other group included linkage of *met-1* to *arg-1* and arg-2 but not to *arg-3*. The *cys-1* and *cys-2* mutants were able to grow on cysteine, methionine, homocysteine, or cystathionine but not on sulfide. Mutations conferring this phenotype are not represented on the current *B. subtillis* chromosomal map (Young and Wilson 1975). The *met-1* mutant has a strict requirement for methionine and may be analogous to *metC* or *metD* of *B. subtilis*. The *arg-1* and *arg-2* mutants were able to grow on arginine, ornithine, or citrulline and may be analogous to *argO* mutants of *B. subtilis*, whereas the *arg-3* mutant, which grew only on arginine, may be analogous to *argA*.

Interestingly, Perlak, Mendelsohn, and Thorne (1979) have isolated a converting/transducing phage for *B. thuringiensis*, designated as TP-13. Mutants that are acrystalliferous and oligosporogenic become crystal positive and spore positive upon infection with TP-13. Thorne has suggested that some prophages of *B. thuringiensis* may exist extrachromosomally. The basis for the conversion is not understood, but naturally occurring prophages may play a role in crystal formation. Strains being investigated in Thorne's laboratory harbor at least three different prophages and at least five plasmids. Tables 15.4, 15.5, and 15.6 show host range, transduction, and cotransduction of transducing phages associated with *B. thuringiensis*. Transduction studies involving transducing phages and auxotrophic mutants also would contribute significantly to an understanding of the molecular biology of *B. thuringiensis*.

CONCLUSION

Entomopathogenic bacteria of pest insect species may offer a unique and relatively safe model system for studying the effects of chromosomal, bacteriophage, and plasmid DNA transformation and recombinant DNA research with procaryotic- and possibly eucaryotic-derived DNA and for

TABLE 15.4——————————————————————
Host Range of *Bacillus thuringiensis* Transducing Phages

Strain	Subspecies	Phage					
		TP-10	TP-11	TP-12	TP-13	CP-51	CP-54
NRRL							
4040	*finitimus*	−	−	−	+	−	+
4041	*alesti*	−	−	−	+	−	+
4042	*sotto*	+	+	+	+	+	+
4043	*dendrolimus*	−	−	−	+	+	+
4044	*kenyae*	+	−	−	+	+	+
4045	*galleriae*	−	−	−	+	+	+
4046	*entomocidus*	−	−	−	+	+	+
4047	*entomocidus-limassol*	−	−	−	+	−	+
4048	*aizawai*	−	−	−	−	+	+
4049	*morrisoni*	−	−	−	+	+	+
4050	*tolworthi*	−	−	−	+	+	+
4055	*kurstaki*	+	−	−	+	+	+
4056	*canadensis*	−	−	−	+	+	+
4057	*subtoxicus*	−	−	+	+	+	+
4058	*darmstadiensis*	+	−	−	+	+	+
4059	*toumanoffi*	−	−	−	+	+	+
4060	*thompsoni*	+	+	+	+	+	+
1715	*thuringiensis*	−	−	−	+	+	+

+ = strain served as host
− = no detectable activity
Source: Compiled by the authors.

genetically manipulating these bacteria. The ultimate goals of such studies are increased pathogenicity, broadened host spectrum, and attainment of in vitro sporulation of *B. popilliae*. It seems evident that the research proposed and in progress would substantially enhance our chances to understand the molecular basis of pathogenicity in these bacteria. There are, however, significant obstacles to be overcome in four areas of endeavor. First, the lack of transformation systems precludes attempts at genetic manipulation by transformation. Second, the isolation, identification, and manipulation of desired DNA fragments alone could easily develop into major efforts. Third, recognition of transformants possessing desirable genes may be difficult to assay. And finally, promising transformed strains constructed by recombinant DNA techniques will have to be exempted from the release prohibition of the NIH guidelines to be tested effectively for efficacy.

It is now difficult to assess the expression of desired genes in *B. popilliae* and *B. thuringiensis* since the experiments are in progress. However, based on the present state of knowledge, especially in regard to the genetics

TABLE 15.5————————————
Transduction of *Bacillus thuringiensis* 4060
with TP-10, -11, -12, and -13

Phage	Recipient Marker	Host for Phage Propagation	Transduction Mixtures		Colonies/ ml
			Cells/ml	PFU/ml	
None	cys-1	—	4×10^8	—	20
TP-10	cys-1	his-1	4×10^8	1.4×10^8	1,740
TP-10	cys-1	his-1	4×10^8	1.4×10^8 (AS)*	25
TP-10	cys-1	cys-1	4×10^8	2.0×10^8	10
TP-11	cys-1	his-1	4×10^8	1.3×10^8	2,130
TP-11	cys-1	his-1	4×10^8	1.3×10^8 (AS)*	20
TP-11	cys-1	cys-1	4×10^8	1.5×10^8	25
TP-12	cys-1	his-1	4×10^8	1.7×10^9	1,040
TP-12	cys-1	his-1	4×10^8	1.7×10^9 (AS)*	30
TP-12	cys-1	cys-1	4×10^8	1.9×10^9	25
None	trp-17	—	5×10^8	—	20
TP-13	trp-17	4060 wild type	5×10^8	1.7×10^8	5,600
TP-13	trp-17	4060 wild type	5×10^8	1.7×10^8 (AS)*	10
TP-13	trp-17	trp-17	5×10^8	3.0×10^8	15
None	his-1	—	4×10^8	—	24
TP-13	his-1	trp-17	4×10^8	3.0×10^8	3,900

PFU = plaque-forming unit.

*Phage (0.1 ml) was incubated with specific phage antiserum (0.1 ml) 15 minutes at 37°C before cells were added.

Source: Compiled by the authors.

and molecular cloning achievements with *E. coli* and *B. subtilis*, we predict with reasonable certainty that tangible results with important implications for insect regulation will be obtained from these experiments.

ADDENDUM

Recently, Martin, Lohr, and Dean (1981) described the development of a method to transform plasmid DNA into protoplasts of *Bacillus thuringiensis* formed by treating the cells with lysozyme and then inducing DNA uptake by polyethylene glycol (PEG). The first application of the PEG-induced protoplast transformation in *Bacillus* was by Chang and Cohen (1979). In their experiments Martin, Lohr, and Dean (1981) introduced a chloramphenicol resistance plasmid, pC194, from *Staphylococcus aureus* into *B. thuringiensis* var *kurstaki* (HD-1). Although they could not isolate this plasmid as a stable extrachromosomal element after transformation, they did obtain expression of chloramphenical resistance and suggested that pC194 acts as an insertion element in the recipient host. Independently

TABLE 15.6——————————————
Cotransduction in *Bacillus thuringiensis*

Markers Tested		Cotransduction with Phage (percent)*				
Donor	Recipient	TP-10	TP-11	TP-12	TP-13	CP-51
his-1	*cys-1*	38	39	74	80	71
his-1	*leu-2*				94	92
leu-2	*his-1*				96	
his-1	*trp-13*	0			31	0
trp-13	*his-1*	0			36	0
trp-13	*lys-11*				22	0
his-1	*lys-11*				96	
leu-2	*trp-13*				14	0

*For each value a minimum of 264 transductants were examined.
Source: Compiled by the authors.

Ryabchenko et al. (1980) transformed protoplasts of *B. thuringiensis* var. *galleria* 69-6 with the plasmid pBC16, a tetracycline resistance plasmid from *B. cereus.* The success of introducing foreign plasmid DNA in protoplasts of *B. thuringiensis* will prove useful in analyzing genetic control of toxin production and sporulation and demonstrates the feasibility of using recombinant DNA in the field of insect pathology.

Also in a recent study Iizuka, Faust, and Travers (in press) examined all 17 serotypes (24 strains) of *B. thuringiensis* for the presence of extrachromosomal DNA by agarose gel electrophoresis. The number of plasmid bands ranged from 1 for *B. thuringiensis* var. *sotto* and var. *thompsoni* to 16 for *B. thuringiensis* var. *kurstaki*. The reported plasmid profiles consisted of both CCC and OC forms, with molecular weights of the CC forms ranging from less than 1 megadalton to greater than 200 megadaltons, depending on the strain. In serologically identical varieties (*thuringiensis* Berliner and *thuringiensis* BA-068, serotype 1; *sotto* and *dendrolimus*, serotype 4a, 4b; *subtoxicus* and *entomocidus*, serotype 6), only serotype 6 strains showed similar extrachromosomal DNA profiles and while a number of strains of *B. thuringiensis* have some extrachromosomal DNA elements in common, distinct differences can be observed in the DNA profiles between serotypes and even among strains of the same serotype.

A concomitant study by the same research team (Iizuka, Faust, and Travers [1981]) compared the plasmid profile of single and multiple crystalliferous strains of *B. thuringiensis* var. *kurstaki*. Plasmid DNA from the multicrystalliferous straing (two to five crystals per cell) was compared with *B. thuringiensis* var. *kurstaki* (HD-1 and HD-73) in an attempt to correlate the presence of plasmids with the production of parasporal crystals and to ascertain whether there is a correlation between plasmid profiles and strains of the same serotype. An 18.62 megadalton plasmid was seen in the multicrystalliferous strain that was not detected in HD-1 or HD-73 strains of *B. thuringiensis*. However, other differences in the plasmid profile made it difficult to associate plasmid DNA with crystal production.

REFERENCES

Ackermann, H. W.; Smirnoff, W. A.; and Bilsky, A. Z. 1974. Structure of two phages of *Bacillus thuringiensis* and *Bacillus cereus*. *Can. J. Microbiol.* 20: 29–33.

Adams, J. C., and Hartman, P. A. 1965. Longevity of *Bacillus thuringiensis* Berlinger in the rumen. *J. Invertebr. Pathol.* 7:245–47.

Afrikian, E. G. 1960. Causal agents of bacterial diseases of the silkworm and the use of antibiotics in their control. *J. Insect Pathol.* 2:299–304.

Andreoli, T. E.; Tieffenberg, M.; and Toteson, D. C. 1967. The effect of valinomycin on the ionic permeability of thin lipid membranes. *J. Gen. Physiol.* 50: 2527–45.

Angus, T. A. 1954. A bacterial toxin paralyzing silkworm larvae. *Nature* 173:545.

——. 1956. Association of toxicity with properties of *Bacillus sotto* toxin Ishiwata. *Can. J. Microbiol.* 2:122–31.

——. 1968. Similarity of effect of valinomycin and *Bacillus thuringiensis* parasporal protein in larvae of *Bombyx mori*. *J. Invertebr. Pathol.* 11:145–46.

——. 1970. Implications of some recent studies of *Bacillus thuringiensis*—a personal purview. In *Proc. IV Intern. Colloquium on Insect Pathology*. College Park, Md., 25–28 August 1970, pp. 183–90.

Angus, T. A., and Heimpel, A. M. 1959. Inhibition of feeding and blood pH changes in lepidopterous larvae infected with crystal forming bacteria. *Can. Entomol.* 91:352–58.

Anonymous. 1977. Petition for exemption from the requirement of a tolerance for usage of the milky spore disease bacterium *Bacillus popilliae* on pastures. USDA, Agricultural Research Service, Beltsville, Md., sec. C.

Anonymous. 1978. Dipel: a biological insecticide. In *Abbott Laboratories technical manual AG 1559*. North Chicago, Ill.: Abbott Laboratories, CAPD, pp. 1–34.

Bailey, L. 1976. The safety of pest-insect pathogens for beneficial insects. In *Microbial control of insects and mites*, ed. H. D. Burges and N. W. Hussey, pp. 491–505. New York: Academic Press.

de Barjac, H. 1978a. Étude cytologique de l'action de *Bacillus thuringiensis* var. *israelensis* sur larves de Moustiques. *C. R. Acad. Sci. Paris.* 286:1629–32.

——. 1978b. Toxicité de *Bacillus thuringiensis* var. *israelensis* pour les larves d'*Aedes aegypti* et d'*Anopheles stephensi*. *C. R. Acad. Sci. Paris.* 286:1175–78.

de Barjac, H.; Sisman, G.; and Cosmao-Dumanoir, V. 1974. Description de 12 bactériophages isolés à partir de *Bacillus thuringiensis*. *C. R. Acad. Sci. Paris.* 279:1939–42.

Beard, R. L. 1956. Two milky diseases of Australian Scarabaeidae. *Can. Entomol.* 88:640–47.

Benz, G., and Altwegg, A. 1975. Safety of *Bacillus thuringiensis* for earthworms. *J. Invertebr. Pathol.* 26:125–26.

Bernhard, K.; Schremph, H.; and Goebel, W. 1978. Bacteriocin and antibiotic resistance plasmids in *Bacillus cereus* and *Bacillus subtilis*. *J. Bacteriol.* 133: 897–903.

Bucher, G. E. 1957. Disease of the larvae of tent caterpillars caused by a spore forming bacteria. *Can. J. Microbiol.* 3:695–709.

——. 1961. Artificial culture of *Clostridium brevifaciens* n. sp. and *C. malacosomae* n. sp., the causes of brachyosis of tent caterpillars. *Can. J. Microbiol.* 7: 641–55.

——. 1963. Nonsporulating bacterial pathogens. In Insect pathology: an advanced treatise, ed. E. A. Steinhaus, vol. 2, pp. 117–47. New York: Academic Press.

Bulla, L. A.; Rhodes, R. A.; and St. Julian, G. 1975. Bacteria as insect pathogens. *Ann. Rev. Microbiol.* 29:163–90.

Carlton, B. C. 1976. Complex plasmid system of *Bacillus megaterium*. In Microbiology – 1976, ed. D. Schlessinger, pp. 394–405. Washington, D.C.: American Society for Microbiology.

Carlton, B. C., and Helinski, D. R. 1969. Heterogenous circular DNA elements in vegetative cultures of *Bacillus megaterium*. *Proc. Nat. Acad. Sci. (USA)* 64:592–99.

Chang, S., and Cohen, S. N. 1979. High frequency transformation of *Bacillus subtilis* protoplasts by plasmid DNA. *Mol. Gen. Genet.* 168:111–15.

Chapman, H. M., and Norris, J. R. 1966. Four new bacteriophages of *Bacillus thuringiensis*. *J. Appl. Bacteriol.* 3:529–35.

Clowes, R. C. 1972. Molecular structure of bacterial plasmids. *Bacteriol. Rev.* 36: 361–405.

Cohen, S. N. 1976. Transposable elements and plasmid evolution. *Nature* (London) 263:731–38.

Colasito, D. J., and Rogoff, M. H. 1969. Characterization of lytic bacteriophages of *Bacillus thuringiensis*. *J. Gen. Virol.* 5:267–74.

Davidson, E. W.; Singer, S.; and Briggs, J. D. 1975. Pathogenesis of *Bacillus sphaericus* strain SSII-1 infections in *Culex pipiens quinquefasciatus* (= *C. pipiens fatigans*) larvae. *J. Invertebr. Pathol.* 25:179–84.

Debabov, V. G.; Azizbekyan, R. R.; Chilabalina, O. I.; Djarhenko, V. V.; Galushka, F. P.; and Belich, R. A. 1977. Isolation and preliminary characterization of extrachromosomal DNA for *Bacillus thuringiensis*. *Genetica* (USSR) 13: 496–500.

Dingman, D. W., and Stahly, D. P. 1979. Plasmids of *Bacillus popilliae*. *Abstract of the Annual Meeting, American Society for Microbiology*, Los Angeles, Calif., 4–8 May 1979.

Domardskii, I. V.; Levadnaya, T. B.; Sitnikov, B. S.; Rassadin, A. S.; and Denisova, T. S. 1976. Transformation of *Bacillus subtilis* by isolated DNA of R plasmids. *Dokl. Akad. Nauk SSSR.* 226:143–46.

Dutky, S. R. 1940. Two new spore-forming bacteria causing milky diseases of Japanese bettle larvae. *J. Agr. Res.* 61:57–68.

Ebersold, H. R.; Luthy, P.; and Mueller, M. 1977. Changes in the fine structure of the gut epithelium of *Pieris brassicae* induced by the δ-endotoxin of *Bacillus thuringiensis*. *Bull. Soc. Entomol. Suisse*. 50:269–76.

Ehrlich, S. D. 1977. Replication and expression of plasmids from *Staphylococcus aureus* in *Bacillus subtilis*. *Proc. Nat. Acad. Sci. (USA)* 74:1680–82.

Ermakova, L. M.; Galushka, F. P.; Strongin, A. Ya.; Sladkova, I. A.; Rebentish, B. A.; Andreeva, M. V.; and Stepanov, V. M. 1978. Plasmids of crystal-forming bacilli and the influence of growth medium composition on their appearance. *J. Gen. Microbiol.* 107:169–71.

Fargette, F., and Grelet, N. 1976. Caractères de mutants de *Bacillus thuringiensis* Berliner, résistant à l'oxytétracycline, ayant gardé ou perdu l'aptitude à sporules. *C. R. Acad. Sci. Paris*. 282:1063–66.

Fargette, F.; Grelet, N.; and Rapoport, G. 1978. Studies on the mechanism of tetracycline resistance in mutants of *Bacillus thuringiensis*. *Biochimie* 60:119–26.

Fargette, F.; Rapoport, G.; and Grelet, N. 1978. Modalités de la résistance à la tétracycline de *Bacillus thuringiensis*. *C. R. Acad. Sci. Paris*. 286:1535–38.

Fast, P. G. 1975. Purification of fragments of *Bacillus thuringiensis* δ-endotoxin from hemolymph of spruce budworm. *Bimonthly Research Notes. Can. For. Ser. Environ. Can.* 31:1–2.

Fast, P. G., and Angus, T. A. 1965. Effects of parasporal inclusions of *Bacillus thuringiensis* var. *sotto* Ishiwata on the permeability of the gut wall of *Bombyx mori* (Linnaeus) larvae. *J. Invertebr. Pathol.* 7:29–32.

Fast, P., and Donaghue, T. 1971. The δ-endotoxin of *Bacillus thuringiensis* II. On the mode of action. *J. Invertebr. Pathol.* 18:135–38.

Fast, P., and Morrison, I. 1972. The endotoxin of *Bacillus thuringiensis* IV. The effect of δ-endotoxin on ion regulation by midgut tissue of *Bombyx mori* larvae. *J. Invertebr. Pathol.* 20:208–11.

Fast, P., and Videnova, E. 1974. The δ-endotoxin of *Bacillus thuringiensis*. V. On the occurrence of endotoxin fragments in hemolymph. *J. Invertebr. Pathol.* 18:135–38.

Faust, R. M. 1975. Toxins of *Bacillus thuringiensis*: mode of action. In *Biological regulation of vectors*, ed. J. D. Briggs, pp. 31–48. Washington, D.C.: HEW. (HEW, NIH, NIAID, Pub. no. 77-1180)

——. 1979. Pathogenic bacteria of pest insects as host-vector systems. *Recombinant DNA Technical Bulletin* (DHEW, NIH pub. 79–99), 2:1–8.

Faust, R. M.; Spizizen, J.; Gage, V.; and Travers, R. S. 1979. Extrachromosomal DNA in *Bacillus thuringiensis* var. *kurstaki*, var. *finitimus*, var. *sotto*, and in *Bacillus popilliae*. *J. Invertebr. Pathol.* 33:233–38.

Faust, R.; Travers, R.; and Hallem, G. 1974. Preliminary investigations on the molecular mode of action of the δ-endotoxin produced by *Bacillus thuringiensis* var. *alesti*. *J. Invertebr. Pathol.* 23:259–61.

Faust, R. M.; Travers, R. S.; and Heimpel, A. M. 1979. Correlation of the biochemical and histological temporal sequence in *Bacillus thuringiensis* δ-endotoxin intoxication. In *Progress in invertebrate pathology*, ed. J. Weiser, pp. 63–64. Prague, Czechoslovakia: Agricultural College Campus Press.

Fisher, R. A., and Rosner, L. 1959. Toxicology of the microbial insecticide thuricide. *J. Agr. Food Chem.* 7:686–88.

Forsberg, C. W.; Henderson, M.; Henry, E.; and Roberts, J. R. 1976. *Bacillus thur-*

ingiensis: its effects on environmental quality. Associate Commitee on Scientific Criteria for Environmental Quality, National Research Council Canada (NRCC no. 15385), Ottawa, Ontario, pp. 134.

Fujiyoshi, N. 1973. Studies on the utilization of spore-forming bacteria for the control of house flies and mosquitoes. In *Research Report of the Seibu Chemical Industry Co., Ltd.*, Special Issue no. 1, pp. 1–37.

Galushka, F. P., and Azizbekyan, R. R. 1977. Investigation of plasmids of strains of different variants of *Bacillus thuringiensis. Dokl. Akad. Nauk SSSR.* 236: 1233–35.

Gryczan, T. J.; Contente, S.; and Dubnau, D. 1978. Characterization of *Staphylococcus aureus* plasmids introduced by transformation into *Bacillus subtilis. J. Bacteriol.* 134:318–29.

Gryczan, T. J., and Dubnau, D. 1978. Construction and properties of chimeric plasmids in *Bacillus subtilis. Proc. Nat. Acad. Sci. (USA)* 75:1428–30.

Hall, I. M.; Arakawa, K. Y.; Dulmage, H. T.; and Correa, J. A. 1977. The pathogenicity of strains of *Bacillus thuringiensis* to larvae of *Aedes* and to *Culex* mosquitoes. *Mosquito News.* 37:246–51.

Hannay, C. L. 1953. Crystalline inclusions in aerobic sporeforming bacteria. *Nature* (London) 172:1004.

Heimpel, A. M. 1955. Investigations of the mode of action of strains of *Bacillus cereus* Frankland and Frankland pathogenic for the larch sawfly, *Pristiphora erichsonii. Can. J. Zool.* 33:311–26.

——. 1967a. A critical review of *Bacillus thuringiensis* var. *thuringiensis* Berliner and other crystalliferous bacteria. *Ann. Rev. Entomol.* 12:287–322.

——. 1967b. A taxonomic key proposed for the species of the crystalliferous bacteria. *J. Invertebr. Pathol.* 9:364–75.

——. 1971. Safety of insect pathogens for man and vertebrates. In *Microbial control of insects and mites*, ed. H. D. Burges and N. W. Hussey, pp. 469–89. New York: Academic Press.

Heimpel, A. M., and Angus, T. A. 1959. The site of action of crystalliferous bacteria in Lepidoptera larvae. *J. Insect Pathol.* 1:152–70.

——. 1963. Diseases caused by certain spore-forming bacteria. In *Insect pathology: an advance treatise*, ed. E. A. Steinhaus, vol. 2, pp. 21–73. New York: Academic Press.

Heimpel, A. M., and Hrubant, G. G. 1973. Medical examination of humans exposed to *Bacillus popilliae* and *Popillia japonica* during production of commercial milky disease spore dust. *Environ. Entomol.* 2:793–95.

Helinski, D. R. 1973. Plasmid determined resistance to antibiotics: molecular properties of R-factors. *Ann. Rev. Microbiol.* 27:437–70.

Helinski, D. R., and Clewell, D. B. 1971. Circular DNA. *Ann. Rev. Biochem.* 40: 899–941.

Hoopingarner, R., and Materu, M. E. 1964. Toxicology and histopathology of *Bacillus thuringiensis* Berliner in *Galleria mellonella* (L.) *J. Insect Pathol.* 6:26–39.

Ignoffo, C. M. 1963. Sensitivity spectrum of *Bacillus thuringiensis* var. *thuringiensis* Berliner to antibiotics, sulfonamides, and other substances. *J. Insect Pathol.* 5:395–97.

—— . 1973. Effects of entomopathogens on vertebrates. *Annals N. Y. Acad. Sciences* 217:141–64.

Iizuka, T.; Faust, R. M., and Travers, R. S. 1981. Comparative profiles of extra-chromosomal DNA in single and multiple crystalliferous strains of *Bacillus thuringiensis* var. *kurstaki. J. Fac. Agr. Hokkaido Univ.* 60:143–51.

—— . In press. Isolation and partial characterization of extrachromosomal DNA from serotypes of *Bacillus thuringiensis* pathogenic to lepidopteran and dipteran larvae by agarose gel electrophoresis. *J. Sericult. Sci. Japan.*

Keggins, K. M., Lovett, P. S.; and Duvall, E. J. 1978. Molecular cloning of genetically active fragments of *Bacillus* DNA in *Bacillus subtilis* and properties of the vector plasmid pUB110. *Proc. Nat. Acad. Sci. (USA)* 75:1423–27.

Kellen, W. R.; Clark, T. B.; Lindegren, J. E.; Ho, B. C.; Rogoff, M. H.; and Singer, S. 1965. *Bacillus sphaericus* Neide as a pathogen of mosquitoes. *J. Invertebr. Pathol.* 7:422–48.

Krieg, A. 1969. *In vitro* determination of *Bacillus thuringiensis, Bacillus cereus*, and related bacilli. *J. Invertebr. Pathol.* 15:313–20.

—— . 1970. Thuricin, a bacteriocin produced by *Bacillus thuringiensis. J. Invertebr. Pathol.* 15:291.

Krieg, A., and Franz, J. M. 1959. Versuche zur Bekampfung von Wachsmotten mittels Bakteriose. *Naturwissenschaften* 46:22–23.

Lamanna, C., and Jones, L. 1963. Lethality for mice of vegetative and spore forms of *Bacillus cereus* and *Bacillus cereus*-like insect pathogens injected intraperitonially and subcutaneously. *J. Bact.* 85:532–35.

Lemoigne, M. 1956. Essais d'utilization de *Bacillus thuringiensis* Berliner contre *Pieris brassicae* (L.). *Entomophaga* 1:19–34.

Louloudes, S. J., and Heimpel, A. M. 1969. Mode of action of *Bacillus thuringiensis* toxic crystals in larvae of silkworm, *Bombyx mori. J. Invertebr. Pathol.* 14:375–80.

Lovett, P. S. 1973. Plasmids in *Bacillus pumilus* and the enhanced sporulation of plasmid-negative variants. *J. Bacteriol.* 115:291–98.

Lovett, P. S. and Bramucci, M. G. 1974. Biochemical studies of two *Bacillus pumilus* plasmids. *J. Bacteriol.* 120:488–94.

—— . 1975. Plasmid deoxyribonucleic acid in *Bacillus subtilis* and *Bacillus pumilus. J. Bacteriol.* 124:484–90.

Lovett, P. S., and Burdick, B. D. 1973. Cryptic plasmid in *Bacillus pumilus* ATCC 7065. *Biochem. Biophys. Res. Commun.* 54:365–70.

Lovett, P. S.; Duvall, E. J.; and Keggins, K. M. 1976. *Bacillus pumilus* plasmid pPL10: properties and insertion into *Bacillus subtilis* 168 by transformation. *J. Bacteriol.* 127:817–28.

Martin, P. A. W.; Lohr, J. R., and Dean, D. H. 1981. Transformation of *Bacillus thuringiensis* protoplasts by plasmid deoxyribonucleic acid. *J. Bacteriol.* 145:980–83.

Miteva, V. I. 1978. Isolation of plasmid DNA from various strains of *Bacillus thuringiensis* and *Bacillus cereus. C. R. Acad. Sci. Bulgaria.* 31:913–16.

Norris, J. R. 1961. Bacteriophages of *Bacillus cereus* and of crystal-forming insect pathogens related to *B. cereus. J. Gen. Microbiol.* 26:167–73.

Perlak, F. J.; Mendelsohn, C. L.; and Thorne, C. B. 1979. Converting bacteriophage for sporulation and crystal formation in *Bacillus thuringiensis*. *J. Bacteriol.* 140:699–706.

Prasertphon, S.; Areekul, P.; and Tanada, Y. 1973. Sporulation of *Bacillus thuringiensis* in host cadavers. *J. Invertebr. Pathol.* 21:205–7.

Reeves, E. L., and Garcia, C. 1971. Pathogenicity of bicrystalliferous *Bacillus* isolate for *Aedes aegypti* and other aedine mosquito larvae. In *Proc. IV Intern. Colloquium on Insect Pathology*, College Park, Md., 25–28 August 1970, pp. 219–28.

Ryabchenko, N. F.; Bukaov, N. O.; Sakanian, V.; and Alikhanian, S. I. 1980. Transformation of protoplasts of *Bacillus thuringiensis* var *galleriae* 69-6 by plasmid pBC16. *Dokl. Akad. Nauk. SSSR.* 253:729–32.

Sebesta, K., and Horska, K. 1968. Inhibition of DNA-dependent RNA polymerase by the exotoxin of *Bacillus thuringiensis* var. *gelechiae*. *Biochem. Biophys. Acta.* 169:281–82.

Sikorowski, P. P., and Madison, C. H. 1968. The effects of *Bacillus sphaericus*, strain 8b, on the Clear Lake gnat, *Chaoborus astictopus*. *J. Invertebr. Pathol.* 10:426–28.

Singer, S. 1973. Insecticidal activity of recent bacterial isolates and their toxins against mosquito larvae. *Nature* (London) 244:110–11.

——— . 1974. Entomogenous bacilli against mosquito larvae. *Dev. Ind. Microbiol.* 15:187–94.

Smirnoff, W. A., and MacLeod, C. F. 1961. Study of the survival of *Bacillus thuringiensis* var. *thuringiensis* Berliner in the digestive tracts and in feces of a small mammal and birds. *J. Insect Pathol.* 3:266–70.

Stahly, D. P.; Dingman, D. W.; Bulla, L. A.; and Aronson, A. I. 1978. Possible origin and function of the parasporal crystals in *Bacillus thuringiensis*. *Biochem. Biophys. Res. Commun.* 84:581–88.

Stahly, D. P.; Dingman, D. W.; Irgens, R. L.; Field, C. C.; Feiss, M. G.; and Smith, G. L. 1978. Multiple extrachromosomal deoxyribonucleic acid molecules in *Bacillus thuringiensis*. *FEMS Lett.* 3:139.

Steinhaus, E. A. 1951. Possible use of *Bacillus thuringiensis* Berliner as an aid in the biological control of the alfalfa caterpillar. *Hilgardia* 20:359–81.

Sutter, G. R., and Raun, E. S. 1967. Histopathology of European corn borer larvae treated with *Bacillus thuringiensis*. *J. Invertebr. Pathol.* 9:90–103.

Tanaka, T., and Koshikawa, T. 1977. Isolation and characterization of four types of plasmids from *Bacillus subtilis* (natto). *J. Bacteriol.* 131:699–701.

Tanaka, T.; Kuroda, M.; and Sakaguchi, K. 1977. Isolation and characterization of four plasmids from *Bacillus subtilis*. *J. Bacteriol.* 129:1487–94.

Thorne, C. B. 1978. Transduction in *Bacillus thuringiensis*. *Appl. Environ. Microbiol.* 35:1109–15.

Toteson, D. C.; Cook, P.; Andreoli, T.; and Tieffenberg, M. 1967. The effect of valinomycin on potassium and sodium permeability of HK and LK sheep red cells. *J. Gen. Physiol.* 50:2513–25.

Travers, R. S.; Faust, R. M.; and Reichelderfer, C. F. 1976. Effects of *Bacillus thuringiensis* var. *kurstaki* δ-endotoxin on isolated lepidopteran mitochondria. *J. Invertebr. Pathol.* 28:351–56.

Van Tassell, R. L., and Yousten, A. A. 1976. Response of *Bacillus thuringiensis* to bacteriophage CP-51. *Can. J. Microbiol.* 22:583–86.

Wille, H. 1956. *Bacillus fribourgensis*, n. sp. Erreger einer "milky disease" im Engerling von *Melolontha melolontha* L. Mitt. *Schweiz. Entomol. Ges.* 29:271–82.

Yoder, P. E., and Nelson, E. L. 1960. Bacteriophage for *Bacillus thuringiensis* Berliner and *Bacillus anthracis* Cohn. *J. Insect Pathol.* 2:198–200.

Young, F. E. and Wilson, G. A. 1975. Chromosomal map of *Bacillus subtilis*. In *Spores VI*, ed. P. Gerhardt, R. N. Costilow, and H. L. Sadoff, pp. 596–614. Washington, D.C.: American Society for Microbiology.

Young, F. E., and Wilson, G. A. 1978. Development of *Bacillus subtilis* as a cloning system. In *Genetic engineering*, ed. A. M. Chakrabarty, pp. 145–64. West Palm Beach, Fla.: CRC Press.

Young, I. E., and Fitz-James, P. C. 1959. Chemical and morphological studies of bacterial spore formation. II. Spore and parasporal protein formation in *Bacillus cereus* var. *alesti. J. Biophys. Biochem. Cytol.* 6:483–98.

16

RECENT DEVELOPMENTS:

MOLECULAR CLONING OF FOREIGN DNA IN PLANTS USING CAULIFLOWER MOSAIC VIRUS AS RECOMBINANT VECTOR

R. J. Shepherd
B. Gronenborn
R. Gardner
S. D. Daubert

Two sites have recently been identified at which new DNA sequences can be inserted into the genome of cauliflower mosaic virus (CaMV) without impairing its ability to replicate in plants (Gronenborn, Gardner, and Shepherd, in press; Daubert, Gardner, and Shepherd, in preparation). The CaMV genome, after propagation in bacteria by attachment to a bacterial cloning vector, was opened at unique restriction endonuclease sites for insertion of the foreign DNA. After further amplification in bacteria and removal of the cloning vector, the DNA was mechanically inoculated to plants. Infection and systemic development of the virus was not impaired by insertion of up to 250 bp of foreign DNA.

Bacterial DNA segments of 60 and 250 bp have been inserted at the unique *Xho*I site of pCaMV10, an infectious clone of CaMV strain CM1841. After mechanical inoculation, replication and cell-to-cell movement appeared to occur in the usual manner in *Brassica campestris* (turnip, variety "Just Right").

Retention of the foreign DNA during the replication cycle in plants was demonstrated by several means. Virus has been isolated from plants, the DNA extracted as described by Gardner and Shepherd (1980) and the pattern of DNA restriction fragments on agarose gels assessed. The array of

255

fragments was found to duplicate that of the original inoculum applied to the plants. The bacterial DNA inserts were also confirmed by Southern blot analysis. Finally, in the case of the 60 bp insert, the DNA was sequenced to positively identify the insert (Gronenborn, Gardner, and Shepherd, in press).

Larger inserts in CM1841 were not retained *in toto.* When viral DNAs with an added 500 or 1200 bp were inoculated to plants, the onset of systemic symptoms was delayed several weeks beyond the usual two-week period. When these plants were analyzed by the rapid mini-screen procedure (Gardner and Shepherd 1980), it was found that the viral DNA had suffered deletions that removed part or all of the inserted bacterial fragment. This observation suggests that only a limited amount of DNA can be accommodated in the viral capsid and that such encapsidation is required prior to the development of systemic infection.

It seems likely that viral DNA with large inserts may replicate as a whole only in those cells that become infected following mechanical inoculation. Plant nuclei are a probable site where viral DNA replicates extrachromosomally as a plasmid (Ansa, Bowyer, and Shepherd, in preparation; Guilfoyle, personal communication). Spontaneous deletions during replication of the plasmid may give rise to smaller genomes that encapsidate successfully. Only then would cell-to-cell movement and systemic infections occur.

Present information indicates the size limit for encapsidation of CaMV DNA is about 300–400 bp beyond the present size of the genome (8030 bp). However, viable strains are known with sizable deletions. These genomes might allow considerably more DNA to be inserted. Comparison of the sequence of undeleted strain CM1841 (Gardner et al. 1981) with its deletion mutant, CM4-184 (Howarth et al., in press), shows that 421 bp of the total of 479 bp of region II of the genome is missing in the latter strain. Such strains may accommodate up to 700–800 bp of foreign DNA.

Foreign DNA can probably be inserted at almost any site in the CaMV genome that is not essential for replication. The fact that about 90 percent of region II can be eliminated without any apparent effect on replication, encapsidation, or cell-to-cell movement suggests this entire gene is dispensable. The *Xho*I site used for insertions in pCaMV10 is in this region of the genome.

In the near future CaMV promoter sequences will probably be identified and become available as bacterial clones. It will then be feasible to insert foreign genes accompanied by an appropriate promoter sequence to seek gene expression in plants.

A second region in which insertions or deletions might be expected to be innocuous is the large intergenic region. This region, of about 1100 bp, probably is not translated (Franck et al. 1980; Gardner et al. 1981). In fact a second insertion site has been discovered in this region. In this case a unique restriction endonuclease site has been created by the introduction of a *Sal* I site in an infectious clone of CaMV cabbage B strain (pCaMV12). Bacterial DNA insertions of 60 and 260 bp have been made into this site with successful propagation of the recombinant DNA in plants (Daubert, Gardner, and Shepherd, in preparation).

REFERENCES

Franck, A.; Guilley, H.; Gonard, G.; Richards, K.; and Hirth, L. 1980. Nucleotide sequence of cauliflower mosaic virus. *Cell* 21:285–294.

Gardner, R. C., and Shepherd, R. J. 1980. A procedure for rapid isolation and analysis of cauliflower mosaic virus DNA. *Virology* 106:159–161.

Gardner, R. C.; Howarth, A. J.; Hahn, P.; Brown-Luedi, M.; Shepherd, R. J.; and Messing, J. 1981. The complete nucleotide sequence of an infectious clone of cauliflower mosaic virus by M13mp7 shotgun sequencing. *Nucl. Acids Res.* 9:2871–2878.

Gronenborn, B.; Gardner, R. C.; and Shepherd, R. J. In press. Cloning of foreign DNA in plants using cauliflower mosaic virus as vector. *Nature.*

Howarth, A. J.; Gardner, R. C.; Messing, J.; and Shepherd, R. J. In press. Nucleotide sequence of naturally occurring deletion mutants of cauliflower mosaic virus. *Virology.*

SUBJECT INDEX

255–57; aphid transmission factor, 100; deletions in genome of, 103, 106, 256; DNA of, 101; encapsidation of, 1, 256; inclusion bodies, 100; infectivity of recombinants, 107, 255–56; insertions in, 255–56; nucleotide sequence of, 107; open reading frames, 107, 256; restriction maps of, 101, 103, 105; RNA transcripts, 105–7; single strand gaps in, 103; strains of, 101–3, 107
caulimoviruses, 85, 100
cell walls, reformation of, 1
centromeres of yeast, 197–98
cesium chloride, 43, 44, 45
cetyltrimethylammonium bromide (CTAB), 40
Charon 4A, 48–57, 121, 123
chlorate, 14
chloris striate mosaic virus (CSMV), 88, 90, 92, 93
chlorophyl deficient mutants, 22
chloroplast, 8, 23
chromate, 13
chromosomes: and genetic transformation, 1; and gene transfer, 21–22; isolation of, techniques for, 12; satellited centromeres of, 12; segregation of, 20–21; of yeast, 195 (see also deoxyribonucleic acid)
ClaI, 65
Cloaca, 228
cloning, 39–58, 63–70, 121, 124, 181, 187–92; DNA vectors for, 39, 40, 45–48, 63–70, 99–107, 126, 145–61, 164–75, 187, 191, 193–96, 198, 203–221, 241–43; 255–57 [binary vehicle system for, 145–52, 181]; in Bacillus, 243; in CaMV, 255–57; of CaMV, 107; isolation of cloned genes, 188–94; of leghemoglobin genes, 62–71; of plant genomes in Charon 4A, 47–48; of Pseudomonas genes, 155, 173–81; of Rhizobium genes, 137, 150; of sequences coding for zein, 73–83; of symbiosis genes, 137–39; of Ti-plasmid, 113–14; [of junction fragments, 121]; in yeast, 187–99
Clostridium brevifaciens, 228
Clostridium malacosomae, 228
colchicine, 13
Coleoptera, 230, 233
colony hybridization, 75–77, 188
colored variants, in carrot, 17
conjugation in bacteria, 134, 164, 168 (see also matings, mobilization, plasmids)
crown gall disease, 112–13
Cycloheximide, 13, 16, 17, 19, 20, 27, 28
cytokinin, 14
cytolysis, 216, 217

dahlia mosaic virus, 100
Datura, 11
Daucus capillifolius, 23
Daucus carota, 13, 23
3, 4-dehydro-DL-proline, 13
deoxyribonucleic acid (DNA), 1, 21–22; Charon 4A, 48–55; and cloning, 39, 40, 45–48, 63–70, 145–61, 164–75, 187, 191, 193–96, 198 [binary vehicle system for, 145–52; sequence coding for zein proteins, 73–83; and symbiosis, 137–39]; common region of, 113, 116; complementary clones (cDNA), 64, 73, 212 [preparation of, 74–78, 80, 81, 82]; donor, 134, 135, 136; and entomopathogenic bacteria, 225–48; expression in prokaryotes/eukaryotes, 192, 203; and genomic libraries, 40, 49–58, 212–15; hybridization of, 45, 48, 67–68, 70, 76, 78–81, 121, 124, 136–37; isolation of, 41–45; passenger, 220; single-stranded (ssDNA) in viruses, 85–96 [and double-stranded (dsDNA), 85]; transferred plasmid (T-DNA), 112–14 [core, 116–18; extent of, 115–16; integration of, 120–24; molecular biology of, 125–26; in octopine tumor genomes, 114–15; right-hand, 118–19; subcellular location of, 124–25]; uptake of in cell, 2, 4, 6; viral, as gene vector for higher plants, 99–107
5′deoxyribonucleotides, 104
Department of Agriculture, U.S., 235
Department of Agriculture, Canadian, 236
Dextran sulphate, 68
2, 4-dichlorophenoxyacetic acid (2, 4-D), 12, 13
diethylprocarbonate, 49
differentiation in bacteroids, 134
dihydrodipicolinic acid synthetase, 15
dihydrofolate reductase, 204
dimethyl sulfoxide (DMSO), 18
Diptera, 232, 233
DNA (see deoxyribonucleic acid)
dominant trait, 15
Drosophila, 39, 218

EcoRI, 48–49, 50, 52, 55, 58, 65, 68, 69, 70, 79, 80, 81, 103, 116, 121, 135, 136, 137, 147, 155, 156, 157, 158, 161, 171, 172, 173, 213, 243
Elcar, 217
electrophoresis, 27, 49, 74, 77, 88, 101, 114, 121, 136, 156, 173, 178, 181, 247
embryo differentiation, 23
embryogenesis, 12, 16, 24–28
embryology, 12
embryonic proteins (E), 27, 28

NAME INDEX

Abelson, J., 189
Ackermann, H.W., 244
Adams, J.C., 237
Adams, J.R., 206, 209, 211
Afrikian, E.G., 242
Ahuja, M.R., 11
Air, G.M., 94
Al Ani, R., 100
Altwegg, A., 237
Alwine, J.C., 78, 106
Amarger, N., 131
Anderson, D.L., 213
Anderson, R.S., 41
Androli, T.E., 232
Angus, T.A., 230, 232, 233
Aoki, S., 2
Areekul, P., 237
Armstrong, T.Z., 26, 27
Arnott, H.J., 215
Ashton, N.W., 16
Ausubel, F.M., 129–43, 136, 176
Avery, O.T., 1
Aviv, D., 20
Axel, R., 73
Azizbekyan, R.R., 239, 241

Bagdasarian, M., 161, 164, 167, 168, 171, 173, 177
Bailey, L., 235
Bajszar, G., 24, 26
Barate de Bertalmio, M., 176
Barjac, H. de, 230, 235, 244
Barrell, B.G., 94
Barth., P.T., 147, 167, 168, 171, 177
Baulcombe, D., 64, 68
Bayliss, M.W., 12
Becker, A., 48, 55
Beckman, J., 189
Bedbrook, J.R., 40
Beggs, J.D., 155, 193
Behki, R., 2, 99
Bell, G.I., 188
Belliard, G., 23
Benckeleer, M. de, 22
Bendich, A.J., 41
Benton, W.E., 48, 121, 188, 190
Benz., G., 237
Berg., D.E., 132
Berg, P., 107
Bergersen, F.J., 131

Beringer, J.E., 131, 132, 176
Berlyn, M.B., 14
Bernhard, K., 242
Berns, K.I., 94
Betlach, M.C., 145
Bhojwani, S.S., 1
Bilsky, A.Z., 244
Binding, H., 19, 20
Bird, J., 85, 88
Blakesley, R.W., 93
Blattner, F.R., 48, 49, 52, 115, 121, 145
Blechl, A.E., 45
Bock, K.R., 88, 90, 92, 93
Bogers, R.J., 130
Boistard, P., 167
Bolivar, F., 145, 161, 167, 173
Bomhoff, G., 113
Bonnett, H.T., 21
Borthwick, H.A., 26
Botchan, M., 114, 125
Botstein, D., 175
Boucher, C., 165, 167, 170, 175, 176
Boyd, J.L., 24
Boyer, H.W., 101, 145
Brack, C., 40
Bradley, D.E., 147
Bramacci, M.G., 239, 242
Brandt, A., 74
Braun, A.C., 14, 112
Breathnack, R., 39
Breton, A., 12, 13, 16
Briggs, J.D., 234
Brill, W.J., 137
Brisson, N., 76
Britten, R.J., 26, 40
Broach, J.R., 155, 193, 195, 197
Broach, L.B., 190
Brown, M., 215
Brown, S.E., 129–43
Bruening, G.E., 101
Bruening, H.J., 145
Brunt, A.A., 100
Brüschke, C., 74, 75
Bucher, G.E., 233
Bulla, L.A., 235, 239
Burand, J.P., 213
Burdick, B.D., 239
Burkardt, H.J., 145, 167
Burr, B., 74, 83
Burr, F.A., 74, 83
Butenko, R.G., 18